Gypsy in the Classroom

OTHER BOOKS BY LYN HANCOCK

Love Affair with a Cougar

The Mighty Mackenzie

There's a Raccoon in My Parka

There's a Seal in My Sleeping Bag

Wild Islands/Pacific Wilderness
(WITH DAVID HANCOCK)
REPRINTED UNDER THE TITLE
Vanderhoof—The Town that Wouldn't Wait

Gypsy in the Classroom

Lyn Hancock

EVEREST HOUSE
Publishers
NEW YORK

To Irv Dawson and the staff and students of Monterey School who remember Gypsy

Library of Congress Catalog Card Number: 79-92182
ISBN: 0-089696-085-4

Manufactured in the United States of America

First U.S. Edition AB

Contents

Preface

Over the past several years apes have been much in the news. While their human-like capacity has long been known and exploited in circuses, movies and television commercials, recent studies have made the amazing apes the subject of much scientific research.

Some of the individual animals studied have become almost household names: chimpanzees Washoe, Sara and Lana and the gorilla Koko for their ability to use sign language; Obang the orangutan for his problem-solving and tool-making capacities.

Scientists usually have two goals in mind. First, they want to see how related, in terms of evolutionary history, the various ape species are to human beings today. Second, since these creatures are our closest relatives, it is hoped that by understanding their behaviour we might come to learn something about the history of our own behaviour. (Perhaps I should say prehistory, for what is being supposed is that aspects of ape behaviour may resemble "the way we were" many hundreds of thousands of years ago when in the tropical regions of the old world our ape-like ancestors began their human journey.)

In addition to these formal scientific studies, considerable information on ape behaviour has been gathered from the accounts of people who have attempted to raise apes in the context of human families. In the story that follows, Lyn Hancock adds to that tradition by describing her own attempts to raise the remarkable gibbon named Gypsy.

Many of you may not know what a gibbon is. When we think of an ape, it is usually the chimp or gorilla that springs to mind. Yet the gibbon is also an ape. One reason why it is not as well known as its cousins is because it is the least studied and understood of the apes. Since the scientific literature on gibbons is sparse and accounts of people raising gibbons almost non-existent, Lyn's book is unique and important.

But Lyn did not write the book solely because it deals with a relatively unexplored subject. She has a message to communicate that operates on several levels. Many know from her earlier books of the tremendous rapport she has with many different animal species and this gift is enhanced by her ability to write (and talk) about her experiences. She has given numerous people, including professional scientists, a feeling for the individuality of particular animals and a sense of the ways in which human/animal communication can be developed.

Yet her account is not personal experience alone. As one of her supervising professors in graduate school, I can personally attest to the diligence of her research–on apes in general and gibbons in particular. Moreover, she has the ability to convey relevant information in a clear and engaging manner. As a result her book not only describes, it also explains. When Lyn observes Gypsy behaving in a way that resembles (or differs from) the way we might expect humans to behave in the same situation, she is able to tie these observations to her knowledge of the biological and ecological conditions that produced the gibbon species, thereby giving us a sense of *why* Gypsy might be acting in a particular way.

For those who may accuse Lyn of being anthropomorphic–that is, ascribing human characteristics to non-human animals–may I say that if such is the case, she is in select company. Over one hundred years ago history's greatest naturalist, Charles Darwin, was also one of nature's greatest anthropomorphizers.

Darwin tried to show that humans had certain traits in common with various animal species and of course this produced a violent controversy. But less well known is the fact that he also sought to establish that animals are not machines, that they possess, in varying degrees, the kind of intelligence, problem-solving ability and communication capacity often thought to be uniquely human. Apes were among his favourite examples. Darwin's so-called anthropomorphic approach illuminated our understanding of biology and anthropology to a much greater extent than the works of writers who chose to emphasize pronounced differences between animals and humans.

When Lyn talks about Gypsy in human terms, she is not saying that Gypsy is human. She is merely pointing out that there are significant points of similarity between the two species. However, often we are caught up in delusions over our uniqueness and all-powerful position and we sometimes forget our kinship to other animals and theirs to us. In a personally touching and yet scientifically informed way, Lyn's story of Gypsy inspires our awareness of both these realities.

Dr. Paul Heyer
Department of Communications,
Simon Fraser University, Burnaby, British Columbia

Nobody wants me. (Photo: Bill Townsend)

1
Looking for a Mother

She was only two months but she looked like an old woman. Her thin little body was frail and heavily lined, her tiny face incredibly wrinkled. She lay on a hot-water bottle in a room heated to a stifling 80°F in honour of her tropical origins. A minuscule huddle of long spindly limbs wrapped tightly around her nakedness, a thumb held constantly in her mouth. Shiny, black blobs of eyes stared up from a frightened little face. She was an ugly, black, scrawny, bug-eyed spider. I could scarcely conceal my disgust.

I listened dully, in embarrassment, to the Zoo Director's instructions.

"To keep her alive you'll have to treat her exactly as a human baby. You'll have to keep her at a constant temperature, feed her formula regularly, preferably Similac, wean her to Pablum, Heinz baby food or Gerber's Food for Infants. You'll have to bathe her daily in lukewarm water containing a few drops each of Dettol and detergent. You'll probably want to put her in diapers. And I suggest you get her established with a pediatrician, not a veterinarian, because she is susceptible to human diseases and will have to be treated just like a human. However, the most important thing to remember in raising a primate is to give it the same love, care and security you'd give to your own baby. If you don't, she will die."

Such was my introduction to Gypsy, an orphan gibbon from the Tacoma Zoo whose mother, grown old, rejected her at birth. The keeper found the baby next morning, a tiny, black, naked thing, clinging desperately with one arm to the bars of the cage, the other arm broken, dangling helplessly.

Gypsy's mother was seventeen years old, had given birth before and was to do so again but for some unknown reason she abandoned Gypsy. Only through the round-the-clock efforts of the Zoo Director and his wife, Norm and Moirin Winnick, did Gypsy survive. Moirin cared for the diminutive ape beside her own newborn daughter until the broken arm was healed.

And then when she needed the constant attention of a mother of her own, Gypsy came to me.

I didn't want her. It was an unprecedentedly cold winter in Victoria, a city in British Columbia that prides itself on roses at Christmas and green lawns all year around. But in this winter of 1968-69 temperatures plummeted to record lows, violent storms whipped up snow drifts to the roof and roads were icy and unmanageable.

Conditions were even worse at Island View Beach where I was responsible for the care of the animals that were brought to our research and conservation centre. Only one room in our house could be heated and none was insulated. There was too little water when the pipes froze and too much when they thawed. There were only wet, salty logs to scrounge from the beach and none to buy from the stores. We relied for survival on bulky clothing, a stout old pot-bellied stove and Victoria's reputation for mild winters.

But it was not only the unusual weather that made Gypsy less than welcome. I was still devastated by the recent death of my blind cougar, Tom, with whom I'd had a close association for several years, and I shrank from becoming personally attached to any more animals.

Yet this is easier said than done. My husband, David, was a biologist who believed that people generally become more concerned with conservation of wildlife and its habitat if they are brought into contact with animals, either by direct experience or by vicarious experience through articles, books and films. He also believed that scientists learn more about animals from practical experience than from the pages of textbooks.

I knew from my own experience in the classroom that if children have the opportunity to see wildlife close at hand in constructive circumstances, they are more likely to grow up as conservation-conscious adults.

So, with my teaching salary, we bought an old rambling bungalow with an attached coffee shop on six acres at Island View Beach near Victoria and converted it into the Wildlife Conservation Centre. This was the base for our activities.

Our original purpose had been to study endangered species, like certain bald eagles and peregrine falcons, and to breed them for release in the wild.

However, in the early sixties there were very few facilities in the Pacific Northwest to care for orphaned or unwanted animals. Government agencies do not have the time, money or inclination to deal with such "problem" individuals. Their solution is to kill them upon receipt or pander to popular opinion and release them into "the wild" where in most circumstances they will either be killed by other animals or be driven back into human territory again. Nor do many zoos and aquariums have the facilities and manpower to pay much attention to education, propagation and research, let alone the additional burden of looking after waifs.

As a consequence, David and I were inundated at the Wildlife Conservation Centre by animal dropouts from both wild and captive situations. Of course, it is very rational to get rid of such problems by bopping them on the head or "putting them down," which is the same thing said in a much nicer way. Many people would recommend just that. However, although such a solution saves a lot of bother and heartache, I have always found it quite difficult to do. Trying to preserve life seems so much more ethical than destroying it–from all points of view, including the one of the animal concerned.

During the winter that Gypsy came to live at Island View it seemed that we were inundated by requests to care for unwanted animals as much as we were inundated by rain and snow.

Already living in the compound outside were the eagles and falcons, the seabirds, the grouse and ptarmigan. These were numbered subjects for avicultural and other scientific studies. A few individual personalities were named: like Sam, an Alaskan fur seal; Pierretta, a Canada goose; and Wynken, Blynken and Wobble, a trio of barn owls. These animals had decided to stay with us after recuperating from near death in the wild. Sam was King of the Compound. In spring he herded into his harem any of the animals that roamed free, just as he would have herded his cows on a Pribiloff beach.

Living inside the house were several parrots whose previous owners had found them too hard to handle. Perched on the clothesline when they weren't flying free around the living room, were two kaka parrots, a yellow-headed Amazon parrot and a bar-headed green parrot. Scarlett, an ancient red-and-blue macaw, had her perch atop our defunct TV set. Fortunately for me, Scarlett had been kept in a zoo cage for so long that her wing muscles had atrophied and she could no longer fly. Her handicap gave me only a slight advantage. Naturally averse to women and especially me, Scarlett, with wings outspread and squawking profanities, attacked me at every opportunity. I had to devise a defensive wiggle to avoid her whenever I went in and out of my bedroom.

When Norm and Moirin decided to deliver Gypsy to Island View personally it was as much to have a holiday from the zoo as to give the gibbon the utmost care and attention en route. Little did they know they were to be billeted in a "zoo" for the visit. As our only heated space was the large open area around the pot-bellied stove in the living room, our human guests had to live there with our animal ones. Other house guests that weekend were Hilary Stewart, a Vancouver friend who had been valiantly surviving a Hancock association for many years, and Gordon Cooper, a sophisticated, pipe-smoking Englishman who was thinking of beginning one. Hilary and Gordon occupied the two guest rooms that abutted the living room.

Despite the difficulties with accommodation in the house and the hazards with transportation on the road, everyone had a hilarious time. A sense of humour helped to overcome the lack of heat. One evening I invited half a dozen other guests and we all partook of an elaborate paella valenciana in the living room.

The parrots flew above our heads out of the way but Scarlett who never missed a party swung down from the TV set and waddled over to the menfolk to pick the peas and prawns from their paellas. Scarlett's dinner-table conversation was unmentionable although she sweetly interspersed her strings of four-letter words with "Cuckoo! Cuckoo!"

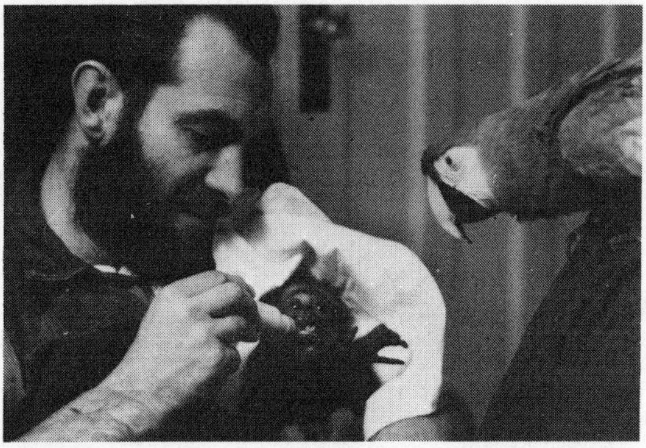

Hallo, what's your name?

The macaw had once lived in the Tacoma Zoo but Norm flatly denied that she had learned her bad language at his place. He assured us that it was all the fault of the little old lady who had donated her to the zoo a few years earlier.

Although Scarlett could breathe sexy love songs when she wanted to, she found she got more attention when she screamed. At one point when our hilarity was at its height, she emitted a triumphant crescendo, high, loud and long. Not to be outdone, the other parrots joined in, all in different tunes and pitches, until the only respite came from turning off the lights and going to bed. That successfully broke up the party.

Meanwhile Gypsy lay quietly on her hot-water bottle in my bedroom beside a space heater. Bug-eyed and babyish, with elongated arms so tightly entwined around herself that she cut gouges in her own skin, the infant gibbon looked terrified. Yet try to comfort her by disentangling her limbs or prying her thumb from her mouth and she screamed in protest. She knew no other security but herself.

She had been born with an instinctive urge to grasp and huddle against something soft and warm and hairy, but no natural mother had responded to her signals. Her subsequent foster mothers like Norm and Moirin Winnick at the Tacoma Zoo and our weekend guests at Island View had satisfied her need for food, but neither they nor inanimate objects like blankets or hot-water bottles provided the continuous comfort she craved. Gypsy needed a full-time mother. In the meantime, without nipples to suck she used her thumbs, without warmth from a hairy body she clung to her own.

It seems ironic now but it took me a long time to tolerate Gypsy. It was not only the weather or other more pressing pursuits that delayed my attachment. She was so incredibly ugly, so embarrassingly human. I remained unmoved by her infant characteristics–her baby face, her clinging, her sucking, her calls–signals provided by nature to bond a mother to her young.

"Watch out, Lynette," warned my mother from Australia, remembering Tom. "Don't get too attached."

"Attached?" I replied scornfully. "Love a spider like that? Warm formula for her? Change diapers? A baby, yes, but this one's an ape."

"Look at it this way," my biologist husband encouraged. "We've had practical experience with seals, otters, bears, raccoons, cougars and all kinds of birds. For a long time I've wanted to learn firsthand about primates. Now's my chance."

As things turned out, it became *my* chance.

Initially I may not have liked Gypsy, but I soon found all kinds of people who did. Despite the hazardous travelling conditions that winter, visitors flocked to the house to see the gibbon, when they hadn't bothered with any of the other animal orphans. The Director of the Provincial Museum brought her a glass-walled terrarium with a false floor that I fashioned into an incubator with a heating pad and electric light bulbs. The local doctor brought his little black book and recorded Gypsy's blood count, body temperature, pulse rate and body measurements. He was interested in comparing the gibbon's development with that of a human baby. The local pathologist came to monitor her blood. My girl friends, all with children of their own, were enthralled by the infant ape and volunteered to baby-sit on weekends.

But what to do with her during the week?

I was teaching English at the time to grades six and seven at Monterey Elementary School in Oak Bay on the Tweed Curtain side of Victoria, almost as far away as one could drive from Island View Beach and still be on the Saanich Peninsula. The previous year I had taught all subjects to a very slow grade six class whose pace quickened considerably under the influence of the various animals, all orphans, that had been introduced into the classroom. Together we had raised an eagle, a red-tailed hawk, a round of robins, a flock of starlings and our three barn owls, Wynken, Blynken and Wobble. The desert tortoises, Ludwig, Gomer and Ponderosa, became the permanent occupants of the class sandbox. As our curriculum was activity-centred and interest-motivated, the animals were an integral part of the learning process which continued outside the classroom when the children and their parents made frequent field trips to Island View Beach.

This year had started a little differently. I was setting up the sandbox on the first day of instruction to accommodate Ludwig, Gomer and Ponderosa when the principal, Irv Dawson, came in to the classroom. He astounded me with, "I don't think we need any animals in school this year, Lyn."

I was speechless, a very unusual condition. After all the rapport that had been built up over the years, both in and out of the classroom, after all the improvement that had been made in the children's writing skills and research methods, and their confidence and sense of responsibility, I found his attitude entirely unexpected and incomprehensible. I could understand if there had been a complaint. But even the janitor had asked to take a tortoise home.

As school is so hectic for the first few months of the new year, I obeyed his wishes and stewed in silence. But as the new class clamoured for the experiences the old class had told them about and as the tortoises looked lonely, I began to make plans to resign. The other teachers were staggered to

Learning is not slow when a tortoise comes to school.

think I would leave but suggested I make sure of another job first. When two were offered closer to home and in schools that cheerfully anticipated Ludwig, Gomer and Ponderosa, I handed the principal my official resignation. For some reason I cannot recall, the confrontation was held in the medical room, which was lucky as it provided a copious quantity of Kleenex.

To my relief it was the principal's turn to be astounded. Crushing the unopened letter and tossing it into the waste paper box, he floored me with a long list of reasons why he refused to let me leave, which I accompanied with two months of repressed tears. He said he had given me an overload of students. He had never met a teacher who came earlier and stayed later. He didn't understand where my energy came from. And the staff had just voted my last year's class "the most eager to learn." As he passed another round of tissues from the nurse's handy supply, he finished by saying, "I must have been out of my mind when I banned animals. I refuse to accept your resignation. From now on you can bring an elephant to school if you want to."

So Gypsy, hardly more than a pound, came to Monterey School and everyone, the principal included, fell in love with

Wynken, the orphaned baby barn owl, under observation.

her. My list of ape-sitters swelled. By the end of the first week everybody in the school had come to see her, the children, the teachers and the parents.

"My, that ape-child has a lot of mothers!" exclaimed the secretary as she wandered into the room with a form to sign.

"And a lot of fathers!" the boys roared in unison.

For the first PTA meeting following Gypsy's admission to Monterey School, there was a record attendance.

"I came tonight really to see Gypsy," confessed one mother. "The children never stop talking about her. We hear of her every night at the dinner table."

"Oh, isn't she cute? Look at those round black shiny eyes. And she's sucking her thumb. Just like a baby," enthused another.

"Jane says she's cute too. But that isn't the right word. She's adorable," commented a third.

One eleven-year-old decided to write a novel called *Guess Who Comes to Our School?* The author, Joan Hollebone, now Joan Isberg with a couple of infants of her own, described Gypsy's arrival in the classroom like this:

> "Is it furry?" asked Andrea.
>
> "Okay, class, I think that's enough questions for today!"
>
> We were to have a surprise the following Monday and everybody was trying to guess what it was.
>
> What a long weekend of waiting! Finally, Monday arrived. Several people ran into the parking area and surrounded the teacher's car. The blue door swung open and two big black eyes stared into the faces of the bubbling, sparkling-eyed children.
>
> "What is it?" whispered one of the girls inquisitively.
>
> "Don't ask me," replied her companion. "Do you know, Bruce?"
>
> Bruce was so shaken by a girl asking him a question, he just shrugged his shoulders and stared at her blankly.
>
> Mrs. Hancock quickly stepped out of her car. She was wearing a clinical mask around her face. Everybody just

Guess who goes to our school? (Photo: *Victoria Times*)

stood around dumbfounded, watching every move she made. They remained silent as they followed her into Monterey School. Whatever was in the box she carried was smothered in towels. All we could see was those big bug eyes.

The twenty people in the classroom stopped what they were doing and crowded over to the table. As the teacher set the box down, a faint murmur was heard from all corners of the room. Then silence.

Just then a loud *oo-oo-oo* sounded from the box and Mrs. Hancock echoed it. She was trying to imitate whatever it was, I guess. Finally, she spoke from behind her mask.

"Now, people, I want you all to stay at least ten feet away from Gypsy."

Evidently Gypsy was its name but we still didn't know what it WAS.

"She is very frail and if she is breathed on she will get germs, so unless you have a mask on, ten-feet clearance please."

"Never mind all the fussy details," one girl whispered to me. "Just tell us what it IS."

The teacher must have overheard her for just then she explained that "Gypsy is a gibbon. G-I-B-B-O-N." She wrote it on the board. "A gibbon is the smallest member of the ape family. The other apes are chimps, orangs and gorillas. Remember, please, that Gypsy is NOT a monkey. Monkeys have tails, apes don't. The so-called Barbary Ape of Gibraltar doesn't have a tail but then that's not an ape, it's a monkey."

The class looked confused. Did it matter? They were more interested in seeing the tiny thing that was all wrapped up in towels.

As if reading their thoughts, Mrs. Hancock continued. "Gypsy comes from the Tacoma Zoo because her mother rejected her. She broke her baby's arm."

From all around was a sound of pity. Some people said "aw" and others said "poor thing."

Finally, Mrs. Hancock lifted the baby up so we could see. The class was open-mouthed, starry-eyed. They just gasped.

Dorcas, a classmate, was appointed official Nanny (or better suited, General).

"Okay, everybody, ten feet without a mask. Don't touch the gibbon, don't breathe on the gibbon. Don't do this. Don't do that. Rule three hundred and forty: sterilize your gloves before handling Gypsy."

General Dorcas took her duties very seriously. People dared not go near the gibbon without their sterile suits on!

The birth of a gibbon in captivity is a relatively rare event. Even rarer is to keep the infant alive. By the time we got Gypsy, half a dozen had been born in Canada but at least two of these had died. Gibbons have notoriously delicate constitutions and are very susceptible to dampness and disease, especially respiratory ones. Hence my insistence on hospital masks.

When the principal one day asked worriedly in the staff room, "What germs does that ape have that your class wears hospital masks?" I replied blithely, "The masks are to protect the ape, not the children."

To keep Gypsy's temperature constant at between 70 and 80 degrees Fahrenheit, I devised two incubators, one at home in the terrarium and another, adapted from a bird cage, at school. Both were glass-walled and heated by two 100-watt light bulbs under a false floor and a heating pad.

For the 17-mile drive by car between incubators, I wrapped her in towels around a hot-water bottle and packed her in a Styrofoam cooler. When she eventually stopped sucking her thumb and started chewing on the green Styrofoam, I substituted a heavy cardboard box. The children always met me in the parking lot and checked to see that she was warmly covered before they carried her away to whatever classroom she was assigned for the day. The towels were for security as well as warmth, a substitute mother to clutch onto and wriggle into whenever she dared unwind her long arms.

The monitor for the day then prepared her formula in a tiny two-inch-long plastic doll's bottle. We fed Gypsy a commercial product called Similac, supposedly the closest known substitute for a human mother's milk, combined with a few drops of multi-vitamins. The formula was then warmed under the hot-water taps in the school washroom. Moirin Winnick first fed Gypsy a quarter of an ounce every hour or two. When the gibbon's appetite increased we put her on a four-hour schedule during which she drank an ounce and a half at each feeding.

Looking after Gypsy soon became routine. Before school she was fed, burped, cleaned and placed in the incubator where she lay all day scarcely moving, mostly asleep. Occasionally she snored. At recess she was again tempted with another bottle and cuddled. At lunch she was fed again, bathed in lukewarm water containing a few drops of Dettol and detergent and dried carefully by the radiator until she fell asleep in the monitor's arms.

For the first month Gypsy revelled in a warm-water soak in a kitchen mixing bowl. Later she developed the natural gibbon antipathy to water, and bathtime became a fight. We could only dip in a section of her body at a time–a leg, an arm, a buttock. The rest stuck leech-like to our sweaters. We were the ones that got washed, not the gibbon.

Joan described it vividly:

> What a chore!
> Squeal, squeal, squeak, squeak, splish-splash!
> Well, there goes the water! Smash! The Dettol bottle.
> Splat! The Ivory Soap!
>
> Boy, Gypsy, are you ever going to get a bawling out from Dork when she comes back. By the time we get you bathed (in about half an hour) you'll be ready for the Mad House... or *we* will.

After school, feeding and cleaning were resumed, cuddling and exercise were encouraged, the bottles and spoons thoroughly washed, the record books maintained, and the hot-water bottle prepared for the journey home.

We nearly killed Gypsy with kindness. We tried to give her too much milk till the doctor told us that barfing in a baby meant an overdose. Then we tried to give her too much heat and not enough humidity. But we did try to keep her clear of germs.

When Gypsy had been with us for two weeks I caught my monthly cold and laryngitis. Now the mother of an ape, I went to the doctor immediately. The patients in the waiting room were enchanted, not with me but with the gibbon.

"Oh, let me hold her," begged one young mother who seemed far more interested in my baby than her own.

"Don't breathe on her," fussed another to her toddler who was trying to compete for attention.

"Mrs. Hancock," announced the nurse. "The doctor will see you now." Was it my imagination? Or did she give the "you" a little more emphasis?

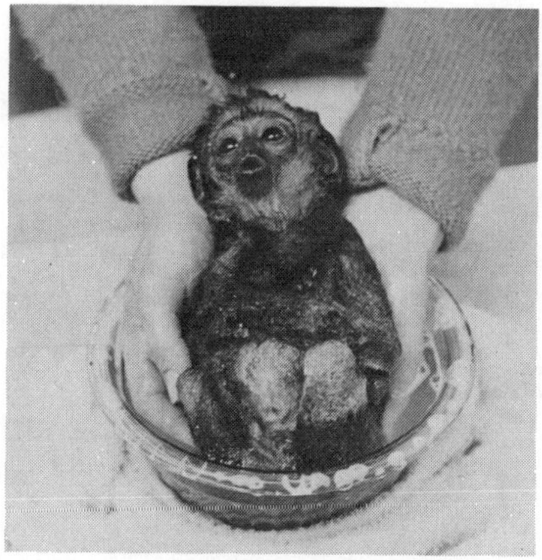

Do I have to take a bath?
(Photo: Bill Townsend)

"We'll look after your baby," volunteered several patients all at once. Gypsy disagreed.

Oo-oo-oo, she whimpered worriedly as I tried to extricate her long arms from around my neck. *Oo-oo-OO-OO-OOOO,* she wailed, and I did too when her fingernails brought blood. Gypsy didn't want to be left alone with strangers, however friendly. Yet at school she didn't even murmur when the children carted her from one classroom to another.

"Mrs. Hancock, I'll see both of you now," declared the doctor from the door of the waiting room. From then on, whether I went for colds or for annual checkups, he always had two patients at the same time. When I went to the druggist to pick up medications, invariably they would be listed not for Mrs. Hancock but for Miss Gypsy McGibbon. He called her his favourite patient.

On that first visit the doctor decreed that I should stay home for a few days of rest and recuperation and that Gypsy should be removed to a germ-free house.

There were germs in abundance at Island View. An elderly gentleman had just left after a bout of malaria and a teenager with flu had taken his place.

I always had difficulty finding assistants who were willing to give the animals the total devotion they required, not only correct care for their physical well-being but activity, diversion and companionship for their mental well-being too. Plenty of people wanted to try but few could stand the pace; and I, being a workaholic with a Capricornian capacity for persistence, was never satisfied. We had just asked one assistant to leave.

He was a dear old man of seventy, polite, aristocratic and well educated, a British-Army type with an ancestry that went back to William the Conqueror. He settled into the guest room, covered the four walls with framed pictures of World War I and spent his first week in bed with malaria. Most of the next few weeks were spent there too. When I discovered a couple of gallons of Mexican gin, rum and tequila missing from the liquor cabinet that is only opened when guests come for dinner, and then when one of the tenants in an adjacent cabin found his wine cupboard raided, I realized what his problem was. It also explained why he wanted to live equidistant from liquor stores and why he spent his pension cheque on taxis.

One day he forgot to close a door and some owls flew away. They must have preferred our company to that of their own as they flew back five days later, but I worried that the peregrine falcons which lived at our Wildlife Conservation Centre under a special permit for breeding might also be lost if Mr. D. was absentminded again.

But how do you put an old man on the street in the middle of winter? Especially a veteran of two world wars! Our assistant had only one topic of conversation, the war, and we had only one, animals, so he came to see for himself that accommodation in the city close to friends at the Silver Threads was really in his best interests, not to mention the animals at Island View. During his stay he had made friends with our cat, one of the domestics, so Tabby went downtown with Mr. D. to keep him company.

As one assistant left with one animal, another arrived with more. Mike boarded his flock of racing pigeons in the compound, then, being unused to the winter dampness of the coast, fell victim to flu and bronchitis and promptly went to the bed so recently vacated by Mr. D.

As hospital masks were deemed insufficient to offset infection at Island View, Gypsy went to our friends the Gyes. Sonya and Michael were enchanted by the new addition to their already large family and were prepared to stay up all night if need be to care for the infant ape.

The first night Gypsy developed shivers and shakes and refused her formula. After much encouragement she accepted it, then threw up and developed a fever. The Gye family doctor made a house call and recommended regular drops of penicillin and a supplementary feeding of warm water, salt, honey, and of all things, coke.

"Things did seem to go better with coke" as Gypsy's temperature dropped, her eyes brightened and she started to *oo-oo-oo* again, this time happily. Like a lot of children, she seemed to prefer coke to Similac. The doctor also suggested she might be suffering from too much heat and too little oxygen. So Michael drilled holes in the bottom of the incubator and experimented with different temperatures and positions of the heating pad and light bulbs.

Gypsy's improvement was short-lived. And we didn't know why. Sonya wrote in the daily diary: "There is something wrong and I don't know what. We are missing something somewhere. I am disgusted and angry at myself that I can't get to the bottom of it. Things are just not right. She behaves like I feel after too hot a bath–blah! What the hell is wrong? Thank God she's taking the coke."

Two days later we found the solution: the right combination of warmth and air. We removed the light bulbs altogether and placed the electric heating pad on the real floor of the incubator, not its false floor, so she had contact only with the towels. She sneezed and coughed for hours afterwards and it was a great temptation not to add more heat. However, even with the heating pad set to medium, we found the correct temperature could still be maintained. And Gypsy slowly improved. Just in time for the Gye household to come down with the flu! With both the classroom and Island View still under siege by bugs, Gypsy went next to my parents-in-law. Mother and Father had warned me against having children ("We will move away if you have a baby") but they didn't seem to mind looking after Gypsy. They even bought a vapourizer for the ape.

By mid-December Gypsy was back home in the classroom by day and at Island View by night. Our own doctor came around again to check and take her measurements. Her heartbeat was 180, over twice as fast as a human's; her body temperature was 99°F, about the same; and she weighed 756 grams, less than two pounds, at two and a half months. She had only increased her weight by 35 grams in three weeks. Obviously this slow response was due to her fragility and the many changes in her life.

A human baby develops even more slowly, generally at one-third the rate of a gibbon. When a human is born, it is completely helpless and can make only feeble movements of its arms and legs. Not until it is one month old can it raise its chin off the ground when it lies on its front. At two months, it can just raise its chest off the ground.

Gypsy at two months would certainly have mastered these skills and more had she been healthy and lived in the wild. But with her rejection at birth, her arm in a cast, her consequent lack of security, it was several weeks before she made any movement at all except for clinging to herself. She screamed shrilly if anyone tried to lift her by her arms. Gradually, as her many mothers and fathers guaranteed food, warmth and, above all, comfort and attention, Gypsy became more active. She started developing into a definite personality.

By Christmas she had stopped sucking and begun to bite

Gypsy, you're just like my baby sister.

the spoon. She lay on my lap flailing all arms and legs, tossing her head, and making little *uh-uh-uh* sounds of annoyance. The only way I could get milk into her tummy was by squirting a few drops whenever she opened her mouth. I feared she'd choke. It was obviously time to add baby food to her formula.

I paraded up and down the supermarket aisle and chose several cases of titles that appealed to me, though I was soon to find out they didn't necessarily appeal to Gypsy. She liked fruit salad, fruit custard, mixed fruit, peaches, tutti-frutti and (where would she get them in the wild?) liver, beef and turkey. She simply hated peas and carrots and mixed vegetables.

As Joan wrote:

On some days there were mixed vegetables from one end of the room to the other. Mixed vegetables in my hair, mixed vegetables everywhere.

Now that Gypsy was eating an average daily consumption of nine ounces, half formula and half canned baby food, she began a steady and striking increase in weight of about 100 grams a month. Her earlier nondescript nakedness was now covered in gradually thickening black fur, dense and shiny on her back, sparse and greyish on her front side, which in the wild would have been pressed against her mother. The parts of her body that were not furred were black and leathery–the palms of her hands, the soles of her feet, the little naked pad on her backside, her diminutive face with its big buggy eyes and protruding ears. Her body was still skinny but heavier at the chest and shoulders than the hips, something like an elongated triangle pointing downwards, a shape that would help her when she started to swing.

When she was four months old a beautiful silvery ruff began to develop around her face and feet.

"Gypsy's going to be *all* grey and silver by the time she's seven," the grade two class informed me one day in the playground.

They must have been referring to a subspecies of gibbon called *Hylobates concolor*, one of whose members lived in an English zoo at the turn of the century. When this gibbon first came to the zoo at six months of age, she was a dark, smoky-grey but soon she turned totally black. Then when she was seven years old and sexually mature, she turned brown and silvery-grey, leaving only a black cap on her head and a few black hairs on her fingers and toes.

Who was Gypsy? And what would *she* become?

In the New Year, with our gibbon healthy, growing fast and beginning to discover our world, Monterey School discovered more about hers.

2
Gibbons in Your Homework

Joan began the second chapter in her novel with:

> I bet you wouldn't believe that such a sweet thing as a baby ape would cause you a month of homework. Well, it will. I was doing my gibbon research report for more than a month.

Actually the children continued their research for the entire eight months that Gypsy remained in the classroom. Being a firm believer that children remember most the interests that are stimulated or developed during their school years, the associations they make while working with peers and teachers, the involvement they feel for their own projects rather than the facts we might try to make them regurgitate, I took advantage of Gypsy's motivating influence in the classroom and the little ape became the centre of the whole curriculum as well as the physical centre of the whole room.

We set up a schedule of infant care, feeding, cleaning and comforting and chose daily monitors to execute it. Everybody in my homeroom class of grade six had a turn, both boys and girls, but I chose one girl in particular to supervise.

It was not till I read Joan's book that I realized there was a complaint.

> The monitor for the day is supposed to be learning something from looking after Gypsy but with General Dorcas patrolling the area no one gets much of a chance to get acquainted with the gibbon.
> Then one day General Dorcas actually got sick.
> Fortunately (or unfortunately) we have a sort of a sergeant-major (someone up there doesn't like us!)–Jackie. Actually Jackie is much better than Dorcas. At least she shares the gibbon.
> Everything was going fine until we overheard Mrs. Hancock talking with another teacher in the corridor.
> "I tell you everything just falls flat without Dorcas. She's a real mother to that ape."
> Of course, Jackie was mad at that statement because she thought herself a better General than Dorcas any day.
> General Dorcas was still away the next day. We broke one bottle and lost the other. We held Gypsy up for her exercises and I guess she couldn't wait as from nowhere it seemed a stream of something watery dropped from the air onto the floor.
> Gypsy strikes again!
> Then to break the silence of everybody's amazement, three or four people burst into hilarious laughter. It was contagious as in two or three seconds the whole room was in hysterics.
> "Just for that, I'll put you into your incubator for the rest of the day!" said the monitor.
> But this punishment was not carried out for long as it was time for Gypsy's bath.
> We were lucky. After her bath there were only two water pails tipped and only three people had to go home to change into dry clothes. (One of whom was me!)
> Although I hated to admit it, WE NEEDED DORCAS!

That's not just a graph–that's Gypsy on paper!

She was away for several days but finally, in the nick of time, she came back.

HOORAY FOR GENERAL DORCAS!

We, I mean she, fixed up the catastrophe in about two hours by rebathing Gypsy and giving her the proper amount of food, cleaning the room, cleaning Gypsy and cleaning up our schedule.

Dorcas, never get sick again!

When caring for something alive, you learn to attend to small details, to maintain routines, to work together for the good of the animal. You may be able to forget your homework or put off cleaning a messy desk. Not so easily can you forget a baby's insistence to be fed or cleaned. I'm sure we all grew in responsibility that year.

We weighed Gypsy's food daily but measured her body weekly. The blackboards at the back of the room were cleared to make way for graphs to plot her progress, and observations were noted in the daily diary, always at the end of the day but also whenever a class member felt there was something to report.

We watched her closely. There were so many things we wanted to know. How did her food intake compare with her gain or loss in weight? How much time did she spend in sleeping and in activity? At what times of the day was she most active? What parts of her body grew longest and most quickly? What did her various calls mean? What did her different types of behaviour signify? How did she eventually expand and explore her space?

Gypsy stimulated every class to write poems, even songs, in her honour, not only the classes assigned to me for English, not only those individuals to whom words came easily, but also those who thought they couldn't write at all.

Joan prefaced her book with a poem:

A gibbon in the classroom
Is something weird I know
But we've got one in our classroom
For everyone to show.

It's been the school's attraction
For people big and wee,
I heard one small girl saying,
"Can we buy one, Mommy?"

But you can't buy them in a pet store
Nor anywhere around
Although we have two in Canada
Southeast Asia's where they're found.

So unless you take a long trip
To Asia far away
You can't have our little gibbon
Cause Gypsy's here to stay.

Gypsy's vulnerability was her most irresistible feature and it was reflected in most of the children's verse.

In Leila Harris':

An unwanted child is what Gypsy is
From her mother's point of view.
She did not care to bother with a baby
And love it too.

She rejected her wholeheartedly
And gave no loving care
She threw her out away from her
Not caring how she'd fare.

Poor Gypsy was then homeless
In a cold dark cage
Until the following morning
The keeper came enraged.

But Gypsy has been transferred
To Mrs. Hancock's class
Where she gives love to many
And where she's loved at last!

In Barbara Dickson's:

Isn't she cute as she lies all alone?
I wish I could have her all my own.
She just lies there with her big black eyes
And as I watch she cries and cries
But as you listen you understand
Gypsy wants a loving hand.

Kept away in isolation
Waiting for some transportation
Right into some arms so warm
Away from everything that could harm
I wish I could have that little dear
To give her love, year after year.

In Marian Brown's:

Tiny squirming wriggling thing,
Hands and feet clasped so tight
Do you still wish for your mother
To cuddle up close to at night?

In Marion Condrashoff's:

A child of the forest dark
And meant to be free
A gibbon came to school
With eyes of misery.
Her fur is soft as silk
She's as gentle as can be
A wandering ape by the name of
  Gypsy!

More than any other animal that I have introduced into the classroom, Gypsy provided a meeting point between animal and man. Any feelings of embarrassment and unease in seeing how close her anatomy, growth and behaviour resembled our own gave way to feelings of awe and curiosity as together we traced the evolution of our common origins. Yes, particularly for me. Sandy Wilson reflected my personal growth in appreciation for Gypsy:

Gypsy is a gibbon so very, very sweet
Gypsy is a gibbon who plays with her feet
When in Mrs. Hancock's arms they both join in chorus
To hear them *oo-oo-oo* is certainly very joyous.

Someone who wasn't joyous was the vice-principal. One lunch hour when Gypsy *oo-oo-oo*-ed to the top of the scale in a resounding triumphant climax, the vice-principal responsible for the discipline of the school stormed into the room.

"And who may I ask is making that ridiculous noise? Stand up and report for detention after school."

The thought of Gypsy leaving the incubator to write lines such as "I'm a naughty gibbon" was too much for her lunchtime companions. A burst of laughter followed, to which Gypsy, surprised by the sudden echo, provided the final crescendo–and only made matters worse. The vice-principal retreated in embarrassment and I avoided him for a week.

As every class from one to seven was involved in learning about Gypsy, we had to scour all the surrounding libraries to provide enough reading material. Between us we borrowed entire primate sections until the bookshelves in our class alone were boosted to over thirty volumes.

We found little factual material. Although chimps, orangs, gorillas and gibbons are our closest relatives, very few scientists had ever studied them.

Five years earlier, Barbara Harrisson had raised orangutans in Borneo. At the same time, George Schaller published his field studies of mountain gorillas from the eastern Congo. Three years were to pass before Jane Goodall popularized her account of living with chimps in Tanzania. In the 1970s Dian Fossey was to study mountain gorillas in the central African republic of Ruanda; Birute Galdikas-Brindamour to observe orangutans in the Indonesian jungles of Borneo; and Penny Patterson at Stanford University in California to hold long conversations with Koko, a seven-year-old lowland gorilla, using more than 375 gestures from the American Sign Language.

But in 1968 when we started to study Gypsy at Monterey School, there was little in the books to teach us about apes as animals. On the other hand, there was much to teach us about apes as humans.

To the ancient Greeks and early Christians, monkeys and apes were vain and greedy. Buttocks were beautiful in those days and monkeys, in not having any, were ugly. Not only were apes and monkeys thought to be physically repulsive, especially in heat when their genitals swelled, but in the eyes of the medieval Church they were morally vile as well. They became symbols of lust, even of the Devil himself.

In Renaissance times, monkeys were finally distinguished from apes and from men, monsters and mythical beings. However, writers until well into the nineteenth century described apes as if they *were* monsters. A naturalist described the gorilla as "exceedingly ferocious." A missionary described it as "hideous, half-man, half-beast, altogether one of the most frightful animals in the world." Edgar Allan Poe wrote of an orangutan that killed a woman with a razor. Edgar Rice Burroughs in *Tarzan and the Apes* combined the traditionally worst features of the gorilla and the orangutan to invent a monster of his own and Hollywood moviemakers expanded on the image. Even in the twentieth century, despite scientific evidence to the contrary, gorillas are depicted in *King Kong* as gigantic, lascivious and uncontrolled monsters.

Joan was to report in her book that on the occasion of Gypsy's six-month-old birthday:

> A couple of wise guys sang:
> Happy birthday to you,
> Happy birthday to you,
> You look like an ape
> And you act like one too!

Myths based on the emotions of fear and disgust die hard. Even when you are trying to change the image!

The gibbon has always had a more likable traditional image than the chimp, orang and gorilla. Perhaps this is because it is smaller, more agile and graceful, and less man-like in appearance than the other apes.

Count de Buffon, who in 1785 was the first to use the word "gibbon," described it as having a tranquil disposition and a delicate constitution.

In folklore, gibbons are often depicted as reincarnated humans. In Burma they were believed to be the re-embodiments of disappointed lovers. In Malaya siamangs which belong to the same genus as gibbons were thought to be descended from a woman who had turned her son away from home and was punished by being transformed into an animal.

In Borneo the story is told of a native who would not kill a gibbon because the *ngarong* (spirit) of his grandfather who died twenty years before was a gibbon.

> When the native was ordered to kill his tabooed animal he first said to it, "I don't want to kill you, but the Tuan who is giving me wages expects me to and the blame is his. If you are really the *ngarong* of my grandfather, make the shot miss you." He then shot and missed three times. On shooting a fourth time he killed a gibbon but not the one he had spoken to.
>
> (from *The Apes* by Vernon Reynolds)

In Siam, now Thailand, researchers have also had difficulty in getting natives to kill gibbons. Interpreters have explained that natives believe that gibbon calls have a beneficial effect on the growth of crops. Rather than shoot them, natives are anxious to keep a large number near their clearings so that the gibbon calls will increase the yield of corn and rice.

The key to our long fascination with monkeys and apes, whatever their image, is how closely they resemble us in appearance and behaviour. The more likenesses we discover and the more we learn to communicate with them, the thinner becomes the veneer of our differences.

When we see apes as physical and mental caricatures of ourselves–pouring tea at a zoo tea party, dressing up, or mixing a gin and tonic–we laugh because it boosts our ego. But let the ape prove himself superior, then we are angered.

Just before Gypsy came to Victoria I was visiting my cougar at the local zoo. Standing in front of the cage opposite was a young man from Newfoundland who was causing much hilarity for the small group of spectators by teasing one of the chimps. He had taken off his shirt and was snapping it through the bars, in and out of the grasp of the chimp. Every time the chimp missed the handout, the crowd roared with laughter. Suddenly the chimp snatched up the shirt and bounded off to the back of the cage. The young man laughed and waited for the animal to return his shirt.

But now was the chimp's turn for fun. He pulled his arm through one of the sleeves and paraded his prize in front of the spectators. The Newfie was first startled then self-conscious and finally angry. But the chimp hadn't finished. Slipping out of the shirt, he swung over to the front of the cage and started snapping the shirt forward through the bars in an exact imitation of his former persecutor. The crowd roared again, a little self-consciously, perhaps out of loyalty for their closer relative.

Time and time again the Newfie stretched forward and closed on the shirt, and time and time again the chimp with split-second coordination flipped the shirt back again. The chimp grinned and the man glowered. The climax came when the man at last got hold of his clothing. His grip tightened, so did that of the chimp, and the shirt ripped in two. In what appeared to be a mocking gesture, the chimp threw him the rest and swung off cackling to the roof of the cage.

"You stupid bloody ape," screamed the Newfie as he made threatening gestures with his fist. Ignoring his compatriots, he could do nothing but swing out through the exit gate–bare from the waist up.

Until very recently, humans have prided themselves that language remained the one unique feature that separated them from the other apes. Several people tried to teach apes to talk but none with any real success.

The most famous of these were Cathy and Keith Hayes who between 1947 and 1954 taught their chimp, Vicki, to voice "Mama," "Papa," "cup" and "up." Despite chimp inability to imitate human speech, Vicki thought of herself as a human rather than as an animal. When looking through a picture book, she bypassed a picture of her true, hairy, unclad father and spontaneously categorized herself by pointing to a picture of Eleanor Roosevelt.

During the last decade chimps and gorillas have been talking to humans and to each other but not by means of human speech sounds. Language is more than just talking. These apes have been conversing fluently by using gestures from the American Sign Language for the Deaf (Ameslan); by placing plastic tokens of different colours and shapes on a magnetized board; or by punching keys on a special typewriter hooked into a computer.

Apes have amazed humans by being able to string words together to make sentences with the words in the proper English order, by asking and answering questions, by giving orders, by teasing, telling lies, describing how they feel and discussing incidents from the past and the future.

Chimp Washoe classified other chimps as "black bugs." She said "Me sad" when her baby died.

Chimp Lucy invented her own words to describe new things. She called radishes "cry hurt food" and watermelons "drink fruit."

Chimp Sara understood "same" and "different", "in" and "on" and "under", "yes" and "no", "big" and "small", "square" and "round". She knew her colours.

Chimp Lana called the orange soft drink "the coke which is orange."

Gorilla Koko created two characters, "a bad gorilla doll" and "a good gorilla doll," and then produced a dialogue between them.

Instead of being impressed by such experiments presently being conducted in several parts of the United States, some scientists took offence that human and animal language abilities were being compared at all. Perhaps they ascribed to the ancient traditions of the Greeks and Hebrews who believed that man's right over the rest of nature was one of dominance and conquest.

But the old barriers are now breaking down. The gap between apes and humans is narrowing. For old words we must devise new definitions and new connotations. Language, in the sense of it being two-way communication, is *not* unique to humans.

Even Scarlett the macaw seems capable of language. Certainly she can imitate human speech (and far better than any ape). But where she approaches a true definition of language is in the canny way she uses different intonations to match the person addressed or a particular situation. When I swat a fly or look intentionally as if to swat her, she yells out, "Stop that!" When I'm most off guard she'll seductively whisper, "Come here, I love you," but only until I get within reach of her beak, then she lunges out with wings outspread and leaves a red gully down my cheek.

If language no longer seems the crucial barrier between man and animals, neither does the use or even the making of tools. Scarlett grooms herself with her beak, as do most birds, but it is also her custom to scratch her head with the shaft of a feather. Sea otters use rocks to pound open abalone shells. Chimps probe sticks into holes or fish for termites with grass stems. They use twigs as toothpicks and nose-pickers, leaves as sponges to mop up rain water or as we use paper towels and toilet paper. Obang, an orangutan, broke off a rock to provide a cutting edge to make a flaking tool similar to the Olduvai choppers of Lower Paleolithic Man.

Neither does man seem unique in having social customs or culture to mark the differences between tribes or nations.

From one monkey in a particular troop in Japan the rest learnt to wash sweet potatoes before eating them. From another monkey the troop learnt to wash the sand off wheat by throwing it into a stream. Each behaviour became a custom that spread through the troop, becoming part of the learned traditions of succeeding generations.

In Tanzania, Scottish scientists described social customs

that were unique to each of three subspecies of chimpanzees. The different groups had not been separated long enough to make genetic differences likely between them. And their foodstuffs and habitats were too close to account for their differences. So the researchers believed that the custom of holding hands while they groomed, which was a unique feature of one particular group of chimps, had been learned the way one ethnic group learns to be different from another ethnic group by transmitting culture from one generation to another.

Until the 1920s and 1930s people had to get their information about apes and monkeys from pet owners or travellers. It wasn't till then that scientific study began in the field and laboratory with the work of Kohts and Kohler, Yerkes and Zuckerman, Schultz and Kellogg. It resurged in the sixties with the work of Schaller and Goodall, and it has continued in the seventies with Brindamour and Fossey.

Thanks to the National Geographic Television Specials, many of us are now familiar with the adventures of Jane and Birute and Dian, the ladies who recently went to the wild to live with chimps and orangs and gorillas. Except for short periods of observation, nobody has gone to live with the gibbons.

But when my class and I scurried into the libraries looking for information on Gypsy's kin and cousins, we found that the one and only ape that *had* been scientifically studied in the field at that time was the gibbon. Dr. Clarence Carpenter's classic work, *A Field Study in Siam of the Behaviour and Social Relations of the Gibbon*, first published in 1940, is today still the main reference on wild gibbons.

Dr. Carpenter, often called the Father of Primatology, led an expedition to Siam in the late thirties. It was officially called the Asiatic Primate Expedition and aptly abbreviated to APE. Dr. Carpenter used scientific methods we take for granted today: he made the same observations over and over again; he supplemented his written descriptions with

Blackboard Jungle (Photo: *Victoria Daily Times*)

stills and movies and tapes; he collected (a euphemism for "killed") entire gibbon groups to check accurately their sex and size.

If Dr. Carpenter was the first to study gibbons in the wild, we were among the first to study gibbons in captivity, certainly the first to do so in a classroom. Research was difficult. We could only scrounge one copy of Dr. Carpenter's long monograph and only one copy of Ernest Hooton's popular summary of it.

Everybody wanted to read the same book at the same time. And when they did they would copy copiously without digesting what they read. So we decided to dramatize the research.

As we began each school day I pretended I was Dr. Carpenter lecturing to his students upon his return from the mountains of Siam. The class decided what they wanted to find out about gibbons, we listed the topics on the blackboard and then, after being formally introduced as the professor, I selected from the scientist's notes the material that would best answer the class's questions.

For the first half of the period the class listened, selected what they thought most important and took notes. For the rest of the time they expanded their outlines into reports using their own words. The skills of research are best begun in the elementary school at a time of highest motivation. When they prepared their own projects on gibbons and then went on to other apes and monkeys, their research was all the better for the practice in listening to "Dr. Carpenter."

During the year we studied Gypsy, another researcher, John Ellefson, had begun to report on the territorial behaviour of gibbons in Malaysia. We were then able to compare his findings with those of Dr. Carpenter and also those of Bernard Riess who had studied a family of gibbons at the Bronx Zoo in New York State. And as our own gibbon grew and developed, we could look for similarities and differences between her and those in the wild.

With the knowledge that so little research had yet been done on gibbons and that their birth and maintenance in captivity were still regarded as rare events, we all felt keen to learn and hopeful that we might find out something that hadn't been discovered before. Of course what we found out about Gypsy in captivity wasn't necessarily normal gibbon behaviour in the wild; but as she had never known another of her own species, we thought we could determine which of her behaviour patterns were innate and which were learned.

Science became alive and important. For eight months at Monterey Elementary it was the very centre of our lives.

3
No, a Gibbon is NOT a Monkey

First we wanted to know what was unique about Gypsy, a gibbon.

Gypsy is a primate, as is man. Primates have similar cells and blood structure. They have stereoscopic vision. They can coordinate their brain, eyes and limbs to reach out and hold objects a short distance from their eyes, to manipulate tools and to judge distance well when they jump and land. Members of primate groups are linked by an elaborate system of communication that is more than they need for actual survival.

Gypsy is also an ape. Apes are characterized further by having short, shallow trunks that do not bend. They have flexible, free-swinging arms that swing their bodies from branch to branch and even rotate them completely as they hang. Apes can also walk upright on the ground, though occasionally non-primates do that too.

Gypsy is not a monkey, though when she outgrew the incubator and got into all sorts of mischief, we often called her one. Myth, as I said, dies hard. Monkeys, except for the so-called Barbary Ape of Gibraltar, have tails. Gypsy instead had ischial callosities. These were naked pads on her buttocks, thickened horny layers of skin upon which she sat.

She shared this characteristic with many of the Old World monkeys but few of the other apes. Gypsy's ischial callosities were hardly noticeable at two and a half months. One month later, at a time when she was becoming increasingly active in the incubator and beginning to hang by her arms without support, they were very well formed.

Gypsy's genus name is *Hylobates* which means aptly "tree walkers" or "tree dwellers." Her species name is *Hylobates lar* or, more commonly, the white-handed gibbon.

Gibbons are the smallest of the apes, probably the most primitive, and yet by far the most successful in numbers and distribution. They live in the tropical rain forests of Java, Borneo, Sumatra, Malaysia, South Burma, west Thailand and southernmost Yunnan.

Opinions differ as to how well gibbons are doing in the wild. In contrast to chimps, orangs and gorillas, they are still numerous. Ellefson in 1973 thought they were thriving in Southeast Asia in the tens, perhaps even the hundreds of thousands. However, the present consensus is that three subspecies are in danger of extinction, a fourth is delicately balanced, and the rest, including Gypsy's subspecies, are rapidly declining because of road building, clear cutting rather than selective logging, clearing of land for crops, and hunting to provide food, animals for research and the pet industry. David Chivers, a recent gibbon researcher, warns that if current trends continue, "we should be prepared for a 96 per cent reduction of the gibbon population" in fifteen to twenty years.

Nevertheless, the gibbon, in terms of evolutionary success, is highly adapted to its natural habitat. Its small body, light weight, long agile limbs and its ability to hang and swing, take it to the very tops and the ends of the trees.

Gibbons are the most arboreal of all the apes. They are the acrobats, the trapeze artists of the animal world. They hurtle through the trees, swinging arm over arm, with the rhythmic smoothness of a pendulum. Brachiation, the name given to

Clingling is as important as breathing. (Photo: Hilary Stewart)

this method of locomotion, was soon to be a household word at Monterey. Though some of the other apes brachiate too, gibbons do it better. Dr. Carpenter described the incredible agility of one gibbon who jumped off a branch which broke beneath her. Instantly, she turned in mid-air, grasped the six-inch stub that was left, swung around it to get on top, then with practically no loss of momentum, made another long outward jump of thirty feet to the next tree.

Another unique feature of the gibbon is its hand. The long, slender fingers are all about the same size but the thumb is much shorter. The hand is like a hook to grab branches and pick hanging fruit. Unlike other apes that have to walk on their knuckles, gibbons can walk on all fours with their palms and fingers flat. They can also run upright more easily than the other apes though in the wild they rarely do.

Of special interest to us in the classroom was the fact that gibbons live in family groups much as we do. The typical gibbon family is a father, mother and their young, usually two, but never more than four. Both parents care for their offspring. Before Dr. Carpenter looked at free-ranging families, zoos commonly caged a large group of gibbons together, an unnatural situation that resulted in fighting and many deaths. It wasn't till zoos displayed their gibbons in small family groups and tried to emulate the conditions of the wild that breeding successes increased.

Gypsy's family was everyone that came into contact with her. It was her infant characteristics that made people want to care for her: her little round head with its cap of black hair, her large bulging eyes, her squashed face and down-pointing nose, her soft elastic skin, her complete dependence.

"Gee, she's just like my baby sister," was a constant cry in the classroom. "Look at the way her fingers cling to mine. See how she snuggles."

To an infant gibbon, clinging is as important as breathing. Its mother may be running along swaying branches, plunging into empty space and bouncing onto a branch. If the baby relaxes its grip for a second it is dead. Gypsy's survival in Tacoma was due to her innate ability to cling to the bars of the cage.

A couple of scientists, Harry and Margaret Harlow who experimented with rhesus monkeys, found that it was more important for an infant to cling than to suck. To show this, they constructed dummy surrogate mothers from wire mesh, covered half of them with terry cloth and left half bare. Half of the terry-cloth mothers were connected to a milk supply, as were half of the wire-mesh mothers.

The Harlows then introduced infant monkeys into a large room which contained several strange objects. The most frightening was a toy teddy bear which advanced across the floor towards them and beat a drum. The infants screamed and covered their hands over their heads. Without exception, they all rushed for comfort to the cloth-covered mothers. Only when they needed food did they choose the nipples that were connected to the milk supply. The greater urge was to grasp and huddle against something warm for pleasure, security and emotional support.

When alone, Gypsy clung to herself or her towel. When picked up, she clung to us. An infant gibbon in the wild clasps its mother firmly around the chest and is supported by her flexed legs. It entwines its fists and feet in the hair of the mother's thick coat. We as mother gibbons had two distinct disadvantages: we were much larger around the middle and our hair was on our heads. So Gypsy climbed up our sweaters and wrapped her long arms tenaciously around our necks. She seemed mesmerized by hair and always fed better when cuddled by men with beards. We decided this was a response to her inheritance, not to the memory of the mother who had rejected her at birth.

Zoos commonly report behaviour aberrations where mothers refuse to care for their young. Some chimp mothers have refused to allow their infants to touch or to cling to them. One gorilla mother pulled her baby from the vagina, bit off a hand, then a foot, and finally punctured its skull with her teeth. A gibbon in the Kyoto Zoo held her baby's head down with her hands and refused to let it suckle. A gibbon at the Metro Toronto Zoo gave birth to several offspring but never cared for any of them. She refused to feed them and put her thumb in their mouths whenever they reached up to her breast. In such cases zoos remove the babies as soon as possible and raise them by hand, as we were doing with Gypsy.

That Gypsy's mother rejected her offspring at birth is strange because she had successfully raised others. At first it was thought that being seventeen years old and barren she might be too old for motherhood. But she did not reject another one later. And old age didn't stop a 29-year-old gibbon giving birth to her ninth offspring at the Adelaide Zoo in South Australia.

Many monkey and ape mothers in captivity have trouble mating, giving birth and rearing their young because they just haven't learned what to do. Nursing is not automatic. The mother must learn how to hold her baby in the right position to find the nipple. The baby plays its part by squealing and squirming till it finds comfort. If either is raised by humans each learns the wrong responses.

Contact comfort leads to love and many mothers have had insufficient maternal affection themselves. The Harlows demonstrated that monkeys raised in total isolation or in sight of other monkeys but without any physical contact became neurotic. They stared into space, clasped their arms, rocked and swayed their bodies for long periods of time, sucked their thumbs, pinched their own skin, made themselves bleed. When they were later brought into contact with other monkeys, they fought savagely and could not copulate. If they had any young, they were indifferent or brutal to them. Such neurotic tendencies were corrected if the infants were allowed to mix with their peers for some 60-90 days after birth. Every added week beyond that time reduced the chance that they could recover from the bad effects of the early deprivation. Playing with peers was even more important than getting good maternal care.

Konrad Lorenz demonstrated that when animals are raised by human beings during particularly sensitive periods, they may not become well-adjusted adults. He proved that if a gosling followed a person shortly after hatching it could then not be persuaded to follow its own mother. He found that such imprinted birds courted humans at maturity rather than their own species even if they had joined a flock of their own species in the meantime. He cited a case where two hand-raised parakeets did mate and incubate their eggs but when they were again exposed to humans they both courted humans and their own bond was broken. Researchers at the Yerkes Regional Primate Research Center showed that if infant chimps were raised with humans they also preferred humans as mates when they reached sexual maturity. In other cases they showed disinterest in sex altogether.

Though evidence is still limited, observers generally agree that human imprinting will not stop gibbons breeding provided they have plenty of gibbon playmates.

What could we do for Gypsy? Certainly, we gave her plenty of love and attention and plenty of playmates, various kinds of furred and feathered friends, not to mention a school full of little naked apes.

I tried to introduce her to other gibbons in zoos, even her own mother, but she acted as if she were terrified of them. The gibbons behind bars were interested in her, but if I held her out towards them, Gypsy would gouge my neck, bury her face in my hair and make little *oo-oo-oo*'s of distress. When I took her to meet her mother at the Tacoma Zoo where she had been born I tried another method. Leaving her with Norm Winnick in the Primate House, I sneaked away outside to another part of the zoo. Norm reported afterwards that as soon as he showed Gypsy to her real mother, she became very worried, turned away and sucked her thumb. Then suddenly she leapt out of his arms, ran out of the Primate House and down the path to find me. I thought of the chimp, Vicki Hayes, categorizing herself with Eleanor Roosevelt and I realized that Gypsy classified herself as a human who lived in a classroom at Monterey School and in a house at Island View Beach. It was the only way she could have survived.

Bimbo was another gibbon who, like Gypsy, was raised by humans and refused to have anything to do with other gibbons. She lived on the grounds of The Prince Royal's College in Chiang Mai in northern Thailand with the college headmaster, Kenneth Wells, and his family. When the Wells had to return to the United States they tried to persuade Bimbo to leave the school grounds and follow the wild gibbons into the forest. She ignored them and answered only to calls from her human foster family. Freedom is relative and Bimbo chose what to us would be imprisonment.

We didn't know what Gypsy's future would be. I had no intention of selling her to a zoo. Except for the last couple of decades, gibbons have had a poor record of living long in captivity. The one notable exception is a white-handed gibbon who lived at the Philadelphia Zoo for thirty-one years seven months until its death in 1938. Strangely enough to us, this particular gibbon spent his life alone in a long, narrow, dark cage. As we lived in North America where she had been born, Gypsy could never be transplanted back to the jungles of Southeast Asia or even introduced to the rainforest jungles of British Columbia. With her rejection of other gibbons and her increasing acceptance of us, we hoped to keep Gypsy part of our family forever.

Her days in the classroom were happy ones. By January when she was four months old she felt secure enough to start exploring. At first she just explored her own body, shaking her arms and legs, clenching and unclenching her hands and feet. Then she crawled on her tummy to the limits of the incubator and stared up at anything or anyone that passed by.

On the way home from school in the car she would pull herself up to a standing position in the Styrofoam box and stare at my driving movements. At night she was fascinated by Scarlett who swung over to tap a conversation to her on the glass roof of the incubator as soon as we entered the living room.

But when there was a strange loud sound or a strange quick movement, Gypsy chittered nervously, flinched, flung her arms around herself, dug her fingernails into her skin or put both thumbs inside her mouth. She could scream her terror in a volume quite disproportionate to her size.

Gypsy's world was still the world of the incubator. But the

children sat beside her and told her about gibbons in the rest of the world, and how they spent their day.

According to Dr. Carpenter, the youngest gibbons wake up first at about six o'clock and a nursing infant gets a liquid breakfast. Then the adults begin their morning calls and arouse other families in the vicinity to answer. As gibbon families are rarely more than a mile away from each other, they are always within hearing distance and often come into frequent contact.

Humans in the shower and gibbons in the treetops both raise their voices in the morning. Such singing may be due to sheer exuberance in starting the day but for gibbons it serves an extra function. The male and the female, each with its characteristic call, conduct a duet to let other families know where they are prior to setting out on the day's business. The male's call is a series of single hoots; the female's great call is a long-drawn-out series of hoots which rise in pitch and power until the crescendo breaks off and dies away in a few diminuendos.

Gypsy, who had never heard another gibbon, made both these calls instinctively though they were probably not fully developed. She also gave little chirps and squeals when she played or wanted us to come nearer, and soft fretting cries when she wanted food.

At about eight o'clock the gibbon family leaves the branches where it has spent the night and swings off along some pathway through the trees to a definite feeding place. If it comes near another family on the move, the males will start calling again. Wild gibbons prefer to battle with their voices rather than with their bodies. If they are competing for a food tree or have met in a no-man's land on the edge of their defended territory, both groups indulge in a vocal battle till one or the other gives ground.

No words can convey adequately the way gibbons sing. Their voices ring out in an ever-increasing crescendo of sound that draws crowds of avid listeners at zoos. If anyone joins in, it always causes a momentary pause in the gibbons' music as they stare curiously at their competitor.

Serious feeding begins about nine o'clock and continues till about noon. Gibbons eat deliberately, plucking food, peeling or shelling it and removing the seeds. As each animal becomes satisfied, the young play about or groom, the mother caresses her infant or rests, and the father either rests or watches for strange gibbons and other disturbances. Everybody then moves to a shady or protected place for about a three-hour siesta of resting and grooming.

By mid-afternoon the family moves slowly out again towards another food tree. After a second period of feeding it returns through the trees to some dense branches where it settles down for the night. Soon after sundown which comes quickly in the tropics, the gibbons sleep for about eleven hours. During a typical day they may have travelled about 2,000 yards.

Family ties in gibbons are strong. When a predator such as a python, a leopard or a human collector strikes, each member of the family may attempt to flee individually through the trees. However, several stories are told of parents returning to look for their stolen offspring and meeting a bullet for their loyalty.

The father of the family not only defends his offspring both in the wild and in captivity, but he also inspects, grooms, handles and plays with it. However, as a young gibbon approaches maturity the relationship changes from congeniality to antagonism. By the time the adolescent is six years old the parents become increasingly intolerant of its presence in the food tree, with the result that it must forage alone on the edge of the family's territory. At the same time it has a strengthening sex drive, an increasing exploratory urge, and all factors eventually combine to force it to leave its family.

It happens in captive families too. Lee Crandall of the Bronx Zoo in New York State described several instances of young gibbons about six years of age persecuted by both parents to such an extent that the family had to be separated.

In the wild, young adults driven from their families meet and form new ones. In some cases siblings of the same group mate and then they separate from the parent group. In others the maturing male may mate with his mother if the dominant male is by some means eliminated. Similarly, the maturing female may mate with her father.

In learning about Gypsy and her relatives in the wild we came to realize not only our physical kinship but our social and psychological links as well. We came to appreciate the values of belonging to a strong family and how its members are mutually interdependent. As we took care of Gypsy we showed love and concern for each other. We were united in a common purpose—her welfare.

I made it!

4
Apes Naked and Not

Gypsy's first snow:
I bet she is wondering what this weird stuff is. It's white and cold, that's for sure, but what is it?
Mrs. Hancock had to stay until four-thirty, waiting for it to stop snowing. We took Gypsy close to the window so that she could watch its movement. She was bewildered.
Finally, it stopped and Mrs. Hancock was ready to go.
"Be sure to wrap her up extra specially warm," she said.
We filled up the water bottle with extra-hot water and thoroughly wrapped Gypsy in towels for her long journey home.
R-r-r-r-r-r-r-r-r R-r-r-r-r-r-r-rrr
Oh, no! The motor won't start.
"Well, you can't leave Gypsy in the cold car, that's for sure." We all agreed. "There is only one thing to do," Mrs. Hancock concluded. "Take her back into the school."
We took her inside the school again and waited. After a long time the car started and Gypsy and the teacher were on their way home.
It snowed again overnight and the next day it was worse than ever. Mrs. Hancock was late for school that morning,

for two reasons. One was the car wouldn't start again and a service man had to come a long way from the gas station to fix it. The second reason was because Mrs. Hancock heard the phone ring just as she was going out the door. She put Gypsy on a nearby table while she answered it. She put the car keys beside the box.

When she came back the keys were gone. She searched high and low but she still couldn't find them. She phoned Mr. Dawson, the Principal of Monterey School, to say she would be late. Again she looked and to get a better look, she pulled a chair away and guess what she found? That's right, THE KEYS.

"Gypsy, you mustn't ever, ever play with my car keys again," she scolded.

They didn't arrive till recess! The car was fixed and so was Gypsy!

(Joan Hollebone. *Guess Who Comes to Our School?* Chapter 9)

I will long remember that terrible winter.

Hilary, who came over from Vancouver to spend part of the disaster season with me, wrote in my diary for December 28, 1968: "Woke up this morning to find everything in the kitchen frozen. Gypsy's feeding spoon was enclosed in a block of ice in a bowl. I thawed it by the stove enough to loosen the ice then smashed it open to get at the spoon. Temperature in her incubator was down to 60°F so I wrapped her in warmed-up sweaters and she heartily ate half a can of custard and one full bottle of formula."

Gypsy was lucky. It was 6°F in our quarters. I dressed like Mother Christmas in parkas, sweaters and spencers (a highly recommended form of thermal underwear from Australia!). Ding Dong the otter had a heated enclosure in the compound outside but when her light bulbs burnt out and I didn't trust new ones to last during the night, she slithered inside the bedroom and crawled under the electric blanket with me.

The tortoises stayed in the classroom sandbox overnight when school was in session but went home with the students during the holidays. Scarlett, happy with her TV set and an occasional spoon of wine, hunched up her feathers and continued to hurl obscenities at the parrots who were perched on a rope rigged above the pot-bellied stove in the living room. The macaw was good at warming ears if not the air.

When the local lake froze over and trapped some of its birds in the ice, I welcomed one coot and two mute swans into my bathroom.

The rest of the house was filled with herring, fortunately dead, but in varying stages of being thawed in the electric stove. Personally, I don't mind herring in the frozen state. It is the warmed-over mush that is hard on the senses. I intended that the seabirds would get it in a condition somewhere in between, but it froze a second time as soon as it hit the outside food dishes and then nobody wanted it, not even the birds.

Unlike mammalian predators who can exist on a feast or famine life style, sea birds are built to eat a little often, so the house soon received more occupants. In came the murres, puffins, oystercatchers, pigeon guillemots and the rhinoceros–no, not what you're thinking, I must be truthful–the rhinoceros auklets.

Things got too much for our assistant so he left the house and also, unfortunately, his pigeons. I resisted the temptation of feeding them to the falcons and so they blithely went ahead to raise more pigeons out in the compound. I left their coop door open but they stubbornly stayed around.

The cold must have drawn Wynken and Blynken together as, to my great surprise, in the middle of winter Wynken laid five eggs. Either Wobble was a rejected suitor or he wanted to see if the rest of the world was just as white and bleak as at Island View Beach because he flew away and never came back. Several weeks later my little white ghost was found dead from exposure in a farmer's field.

The best part of that disastrous winter was the support I had from the staff and students of Monterey Elementary. Irv Dawson kept telling me he could write a book on my exploits and presented me with a gift from the school which I opened in the staff room with everybody watching: it was a large, beautifully wrapped, exquisitely trimmed, white enamel *potty*. After I got over the giggles I used it often, though to empty it I had to don all my heavy outdoor gear and run over to the municipal toilets as the pipes at Island View Beach remained frozen most of the winter.

The teachers were very sympathetic. They covered for me if I got to school late and even when I didn't get to school at all. The children pushed my car when it wouldn't start and everybody volunteered to help me look after the animals.

I took what Irv Dawson called the Chipmunk Route to school, the scenic drive around the Saanich Peninsula, rather than the busy highway. It provided a quiet interlude between the panic at home and the panic at work. It was a long road, winding through leafy lanes and up and down gentle hills, and it became even longer when it snowed.

One day after a fresh snowfall with Gypsy lying beside me as usual on her hot-water bottle, I skidded around a corner into a dozen cars stuck at various angles across one of the hills. If I had time to think, I would have stopped and, as a maiden in distress, waited for a modern-day knight to rescue me. Instead, I pushed on recklessly till I managed to manoeuvre in and out and around each car almost to the top of the hill. Just at the brim I realized my temerity and made the fatal mistake of slowing down. I came to a complete stop skewed sideways and obstructed the path of everyone behind me. My exclamations and Gypsy's echoing screams, not to mention the position of my car, attracted the attention of the man I was blocking and he patiently pushed me over the hump of the hill. Ten miles and an hour later I got to school.

Some days I didn't get out the door. Living at Island View within a few feet of the ocean, our house was constantly obliterated that winter with wave-pelted logs and wind-flung snow. The municipal snow plough cleared the municipal park land beside us and reached the boat ramp a few feet from our door. Those few feet were vital in deciding whether or not my car would make it to school. I would open the front door, push away the drifts and start sweeping and shovelling a path to meet where the municipal man had been. Once, with an eye on the clock and another on Gypsy, I tried to reverse through the drifts. Vain hope! I was stuck within seconds, the car as a unique but unwanted decoration on an otherwise bleak white cake.

Other times I made it out the front door but broke down on the road getting to the Chipmunk Route. With my mechanical abilities on a par with my driving capabilities, I usually abandoned the car; and with Gypsy piping muffled trills from a package of towels and sweaters under my arm, I crunched across the fields to a neighbour to phone a garage.

If I was lucky I broke down in a busier thoroughfare and crept into a garage on my own. I remember once the attendant did a double take to see tiny Gypsy swinging on the steering wheel. He called to the others and marvelled, "Hey, Joe. Hey, Bill. Come here. Look at the monkey."

"It's not a monkey," I murmured automatically, realizing that I was going to be even later for an early-morning concert rehearsal. "It's an ape."

"Well, ape or no ape," he replied with enthusiasm, "she'd make a great grease monkey for us! Why don't you bring her here more often?"

Eventually after three months of misery, things thawed and I only had the cleanup to contend with. Incredibly, Gypsy survived, even thrived through it all.

Everybody was interested in her progress. Helen Vernon on duty in the school yard reported that a shy little girl in grade two came and asked, "How is the giblet today?"

Arlene Mackenzie said that a boy in her class informed her that the giblet in question had just brachiated along Dorcas' arm. "And guess what?" enthused one of the lunchroom monitors. "Gypsy has just eaten her first baby cookie in Mrs. Kellock's library period. She eats just like my baby sister."

Even the long-suffering janitor was impressed. He strung a dog leash of baubles across the top of her incubator and, curiosity now replacing insecurity, Gypsy reached up to investigate it. She had already begun her swinging practice by hanging onto our arms and pushing with her legs against our chests. Once on the way to school, intrigued by the activity in getting the car started, she pulled herself out of her box to a standing position and tried to swing from the steering wheel.

Once a week the children stretched Gypsy out on the table to take her measurements. About the same time she began to swing, they informed me that her arms were growing faster and longer than her legs. Although the length of a gibbon's upper arm is comparable to the other apes, its forearm is greatly elongated to help it in swinging through the trees. With her arms growing longer and stronger, six-month-old Gypsy began to brachiate. Joan described this historic event in the life of a gibbon:

> Gypsy has started to swing. A chain and a rope have been strung from one side of her incubator to the other. Of course, she was curious to see such odd new things. So she sat up and looked at them. Then saying little *oo-oo-oo*'s to herself, she stood and grabbed the rope. She held on and suddenly she swung. She held on with her long scrawny arms and pushed against the glass, then excitedly she swung some more, hand over hand along the rope. She must have thought this was fun as she kept doing it until she got tired. She really loves brachiating and now all day she swings happily.
>
> Gypsy, you're a real swinging chick!

As the weather had improved and she had the protection of her thickening coat, we removed the light bulbs and false floor in her school incubator and gave her more room to brachiate. Her coordination improved daily. She exercised all day within the confines of her incubator. Although she hadn't yet learned to walk, she crawled around on all fours like a little black crab or else she lifted herself up by her arms and hauled herself from one point on the ground to the other.

Her greatest urge was to lever herself up to the highest point possible. She was happiest when climbing on the children's heads or hanging onto their hair. Perhaps this was because she usually lay at the bottom of the incubator and anything that attracted her curiosity was above her. We must have looked like giants. Accustomed to being disturbed from below and not anticipating disturbances from above, wild gibbons direct their gaze around and down but never upwards.

Gypsy learned new movements mainly by trial and error. She would swing from the chain, then, building up momentum, perhaps accidentally swing overhand backwards. Her shrill screams as she experienced something new made us all stop what we were doing. Gypsy accompanied all her new learning experiences with a soft crying which stopped when she achieved confidence.

Zoos, accustomed to the confident, constant activity of baby gibbons and the close attention given them by well-adjusted mothers, nevertheless guard against any accidental falls by cushioning the floors of their cages or adding more cross bars. Our incubator was too small for accidents but it soon became clear that Gypsy's gymnastics required more space.

Anyone associated with animals is sure to have a storage room of crates and boxes. I rustled around ours to select two for Gypsy's new homes. I chose large cages that once housed canaries. They were shaped like a grandfather clock, glass in front and solid behind, though the back had a door which much later Gypsy learned could be lifted up. I scrubbed them thoroughly, dried them in the sun which at long last had returned to Victoria, polished the glass and inserted light bulbs for heat if it got cold again. Jerry, one of the university students who rented our beach cabins, spent a weekend fixing them up inside with four-foot trees. We put one house in the living room beside Scarlett. We drove the other one to school.

Gypsy looked so tiny and vulnerable lying at the bottom of her huge new room that I almost felt like returning her to the old one. For the first day she just lay with her thumb in her mouth, staring upwards. The first limb was only a foot above her head but she made no move towards it. The next day, with much squeaking and *oo-oo-oo*-ing, she pulled herself up to the first branch and swung on it. In the afternoon she managed to sit on it with one arm clinging tightly to the main trunk of the tree and one foot hanging down ready to touch the familiar floor. She graduated to each branch in turn, using the lower one as a springboard to stretch and swing up to the next one. When Jerry and I cheered her on vocally with, "Come on, Gypsy, you can make it," she was startled. Instantly, she swung down to the security of the floor again, her thumb stuck solidly in her mouth. We left her to experiment in silence.

Having a gibbon in the room is a very efficient way of maintaining discipline. When a new class came in to find Gypsy lying quietly and perhaps sucking her thumb, their immediate response as they crept past the incubator was, "Aw! Isn't she sweet?" To which I replied, "Now you do the same thing and I'll think you're sweet!" This brought a laugh but it also ensured a quiet introduction to the next lesson.

Gypsy disliked loud noises. If a class laughed a lot or even talked too excitedly, she chittered with her mouth agape. If the children made a din, she joined her whoops to theirs in a mutual chorus. Sometimes she screamed. Then, as if frightened by her own outpouring, she huddled in the bottom of the tree house, once again with thumb in mouth. It quietened everyone in the room. Nobody wanted to disturb Gypsy–though later on in the year, *she* had no similar qualms.

One Saturday morning while I was lazing late in bed, I awoke suddenly to hear a plop then a terrified scream, accompanied, a second later, by Scarlett squawking raucously, "What's the matter?"

I ran out into the living room to find that Gypsy had fallen from the topmost branch of her new house. The distance was not great and she had fallen safely on the padded floor and soft towels, but she was severely frightened. I took her to bed with me and that soon became a regular weekend morning activity.

And the emphasis was on activity. She refused to lie still. She used the bed as a trampoline, hopping and bouncing and jumping so there was no way I could get to sleep again. Then she climbed over the mounds of bedclothes and pulled at the curtains with all four limbs. Her thoughts were always directed upwards, her long arms always probing challenges above her head. She crawled over my face, heaved herself to the bedstead and fingered the thermostat for the electric blanket. In no time at all, she had learned to dial the numbers.

She was now increasingly interested in handling objects. She soon learned to take the thermometer off its hook in the incubator. As usual she licked it, bit it, curled her long fingers around it and carried it to the top of the tree. Then she put it into her mouth. We decided to do without a thermometer. At the hairdresser's she stood up on my lap and with one hand clinging to the sweater for balance, she tried to reach out with the other to catch the moving comb. Then she tried to clutch at my hair.

Like human infants, she transferred everything to her

mouth. She was like a little vacuum cleaner, sucking up everything in sight. Sometimes Dorcas put her down on the classroom floor, and one day she leaned down and licked up an ant. Our classroom, except at feeding or bathing times, was cleaner than it had ever been before. Even the messiest children watched what they dropped. Nobody wanted Gypsy to swallow anything she shouldn't.

Zoo keepers report that a mother gibbon is very quick to intervene when her infant reaches out to grab an object she deems unsuitable, even if it's a piece of banana. Although a gibbon in the wild is constantly reaching out to grasp leaves, fruits or flowers, nobody has observed any disciplinary action on the part of the mother.

Noting Gypsy's developing interest in objects, each day the children brought her a variety of toys of different sizes, textures and shapes. Sometimes she picked them up with her hands, other times with her feet. Most often she would lean down and pick them up with her mouth.

When she was old enough to drink from a saucer, she did the same thing, putting her face directly into the water to suck. Captive gibbons often do this but it is very disadvantageous for wild ones since it restricts their visual field and leaves them relatively defenceless against predators. More commonly, free-ranging gibbons dip their fists into water that has collected in pools, throw their heads backwards and suck the drops from the hairs on the back of their hands and knuckles. Although she had no other gibbons to imitate, Gypsy used the same methods. Jean Delacour, who for many years kept gibbons at liberty on his Normandy estate at Clères, says this approach most closely resembles that employed by primitive man, another of our links with gibbons.

As the children gave Gypsy objects to pick up, they noticed that although her limbs were very well suited for support and locomotion, they were poor for manipulation. If they placed small objects on a flat surface, an abnormal situation for a wild gibbon, Gypsy would scoop them up by turning her hands sideways so that her little finger rested against the table. Occasionally she picked them up by using her thumb and the side of the second joint of her second finger. She didn't use her fingers independently as we would but rather as a unit like a spade or a hook.

When we looked closely at Gypsy's hand we could see her problem. Her fingers were all equally long and slender but her thumb was very short and separated by a deep cleft from her first finger. There was another deep cleft between her great toe and the outer digits. Such a hand and foot enables a gibbon to hook onto branches and pluck at hanging fruit. But it is very awkward for picking up small objects, as we found out ourselves when we tried to imitate Gypsy's situation. The children pressed their thumbs tightly against the edge of their fingers and tried to pick up a pencil. They were just as clumsy as Gypsy.

Their advantage was in their opposable thumbs. Gypsy's advantage was in being able to use both her hands and feet interchangeably. Soon she could dangle from one arm, swing out with another to pull in hanging objects, transfer them to either or both of her feet, pluck at them with her free hand and carry them to her mouth.

With her increasing coordination and her increased room to practise, it was a sheer delight to watch her many and varied movements. Seldom did she lie on her back as she used to. Now she lay on her slightly butterball tummy and propped herself on her elbows to stare at us. A wild gibbon rarely reclines on its belly. If the branch is large and round enough, it stretches out on its side with head resting on its upper arm and its legs flexed sometimes up to its chin.

With more branches to climb, Gypsy hung by her arms for minutes at a time and bounced her feet on anything within reach. Then she swung arm over arm to the end of the branch, brachiated back again and, as if realizing she was putting on a show, sat with her hands curled around the trunk looking out at us.

As expected, Gypsy was much more active in the familiar tree house incubator than anywhere else. When she was taken out to join the children for lunch at their desks, an unfamiliar place surrounded by much open space, she sprawled flat, spread-eagled across the surface, with her thumb stuck solidly in her mouth again.

As she got used to the routine and the children bribed her with baby cookies, she gained confidence enough to sit up. She held out the cookie at an awkward angle and sometimes dropped it. If she put her mouth down on the floor or desk and tried to pick it up that way, the children yelled "Foul!"

"We must bring her up properly," said Dorcas primly.

Actually, most of the class preferred to feed Gypsy in the incubator. In the manner of wild gibbons she would close her fingers around a cookie and clamber up to the topmost branch to eat it. She was certainly the centre of attention and she enjoyed it.

Gypsy stayed on Similac and Heinz baby foods until she was almost seven months old. I must confess this was more because of my fear that she wouldn't have a rounded diet for healthy growth, as well as an oversupply of cans in my cupboard, than her own preferences. Come to think of it, Gypsy never did like baby food very much.

When she started biting chunks out of the green Styrofoam box on the way to school and they later appeared in her feces, when she insisted on closing her mouth around the spoon that was feeding her, we knew she was teething. At two months she had already sprouted her top and bottom incisors; at five months she had grown almost all of her baby teeth. And as Joan described in her book, Gypsy obviously knew how to use them.

> *Ow-w-ww!* That was the first sound to say that Gypsy was teething. Another was *Ye-owch*! Although her teeth weren't yet full grown and didn't hurt that much, she kept them in for minutes at a time.
>
> Finally we decided that it was high time she had a teething

ring. Dorcas trotted out to the store and bought one.
For a little while this worked fine. After a week or so she absolutely neglected it. So it was back to the drawing board. But no sooner had I reached for the chalk than there was a stifling *A-h-h-h-h-!*
She chewed on anything she could get her nails on, so we decided it was time to put a bit of discipline into her noggin. Every time she bit someone we would turn her lip under her teeth and she would bite herself instead. It may seem cruel to you but it sure solved the problem of Gypsy's teething.

Gypsy certainly had the equipment. Now she had to learn how to use it in socially accepted situations. This meant biting on solids rather than us.

Every lunchtime the children would take her out of her tree house and experiment to find out the kinds of foods she liked. Strangely enough, she spurned bananas and oranges, at least initially, and instead preferred ham and mustard sandwiches. In our nutrition lessons we worked out combinations of food that would help her grow healthy and the children took a greater interest in their own diets. I am sure there were arguments in the mornings as mothers prepared school lunches and children complained, "Don't put that in. Gypsy won't like it."

At six months Gypsy was eating eggs and meat, raw fruits and vegetables, in addition to the daily selection from her lunchtime companions' lunches. Like most children, she preferred sweet things if she could get them–candy, chocolates and sugar; or foods with no nutritional value, like the map she chewed on the way home from school. Gypsy, like her counterparts in the wild, was meticulous in how she examined her food. She removed the pits, small seeds, and even spots of decay. She selected what she liked and spat out the rest.

She knew her own mind. Bath time was already a fight. Often feeding time was too, if she wasn't hungry or if she was more interested in play. With her growing agility she had a definite advantage. She lay on her back, flailed her limbs in our faces, jiggled her head back and forth, and rolled her tongue forward to push out any food she didn't want to eat. Sometimes she put on a regular temper tantrum and added screams to her body oscillations.

Yes, Gypsy, in the security of her classroom family, was growing up healthy and independent. Joan wondered if she might become spoiled.

YOU . . . YOU . . . GIBBON

Gypsy, you ape, you . . . you . . .
Gibbon.
Sitting in your nice warm box,
Climbing up your branchy tree,
Children feeding you morning, noon and night,
Humans giving you such care, love and attention.
Do you like all this?
Or is it a boring routine?
Watching people all around you
Babbling strangely
In a classroom
Where you sit every day
(Except Saturdays and Sundays)
Like a queen being waited on
By servants.
Gypsy, you . . . you . . .
Gibbon.

5
The Giblet at Six Months is a Real Swinging Chick

News spread that we had a gibbon in our classroom and a photographer for the Victoria Times phoned Mrs. Hancock to ask if she would let him take a few pictures for the newspaper. Although the teacher was hoping to do this herself, she accepted and a couple of days later he came to one of our language periods.

Of course everyone wanted to get in the picture, so the photographer told everyone to sit down. He selected three people to be in the picture (of course one of them was Dorcas). They were to be looking at Gypsy as Mrs. Hancock held her. He took several pictures and said the editor would select the best one for the paper.

Mrs. Hancock told us the picture would be in that night. Everyone frantically searched through the evening paper. All arrived back at school with the same question. "Mrs. Hancock, I searched all through the paper last night. Where's the picture?"

"I'm sorry but you'll have to wait until tonight."

That evening everyone tore open the paper again and searched through it thoroughly but still no picture!

"It might be in tonight, class, but don't get your hopes up!" said Mrs. Hancock the next day.

The next evening edition, there was no scramble for the paper as the kids did not expect to see the picture. There it was though, sure enough, on the first page of the second section.

It wasn't a very big article but we were proud and I'm sure any class would be to see such fame as our Gypsy brought to our school. There were pictures in magazine sections of weekend papers right across Canada showing Mrs. Hancock in another school with other pets she had taken to school.

But it was our Gypsy that made the cover.

(Joan Hollebone. *Guess Who Comes to Our School?* Chapter 5)

Immediately after the picture appeared in the local paper, I went to a meeting at the school board and Gypsy, sitting complacently in a cardboard box beside me, charmed everybody as usual.

"I heard from the SPCA, that they had received several phone calls from people wondering if you had the proper certificate to look after Gypsy," one of my teaching colleagues whispered during the proceedings.

"And what kind of a certificate would that be?" I retorted a trifle sarcastically.

Smarting from the memory of somebody who a year earlier complained that I had caused her daughter a traumatic experience when I pickled a dead day-old robin for my science class, I left the meeting to phone Kay Beeching, the wife of the local SPCA Director. She assured me that she knew of no calls, nor had reported any. The rumour remained an unsolved mystery.

Prejudice is often fathered by ignorance. "Oh, she scratches, she must have fleas," said one little girl in disgust. I assigned her, under my supervision, as the next monitor so that when she washed Gypsy she could see for herself that for this gibbon there was no relationship between fleas and scratching. Gypsy was merely grooming herself.

Grooming was instinctive to Gypsy. She would sit in front of me for an hour, mouth open, making little smacking movements with her lips as she picked out minuscule pieces from my sweater. Or she would part my hair to scratch away tiny flakes of dandruff. She used both her hands to separate the strands of hair then held them in the parted position while she removed the most minute particles, a scratch, a scab, a loose hair, a blemish. I had to limit her interest in moles. She chose areas on my body that were, for her, the most difficult to get at–the head, ears and neck, the shoulders, down the centre of the back and over the rump.

Gypsy groomed herself in the same way she did me but more often as she grew older. We groomed her too. While we scratched and caressed her little body, particularly about her head, she became so relaxed she often fell asleep.

Primates groom to keep their fur clean and smooth and free from parasites. But for some primates the social function of grooming is more important than the biological one. The species that have the most clearly defined social hierarchy (where some animals in the group are dominant and others are subordinate) are the ones that do the most grooming. The more aggressive baboons and macaques groom more often than the gentler gibbons and gorillas. Grooming and being groomed are signs of friendliness and sociability. Sometimes an animal of lower status can ward off an attack from one more dominant by facing away and presenting its back for grooming. Such an appeasement gesture serves to overcome the other's aggression.

A gibbon family is small and the parents, being of relatively equal status, alternate between grooming and being groomed. Such reciprocal activity strengthens their attachment and keeps the family together. Sometimes a female grooms the male after he has engaged in a vocal battle with another family. Reassured and morally supported, he then returns to the fray.

When people had direct contact with Gypsy they often lost their prejudices. My parents were disadvantaged by distance. Every few months I sent them a film of Gypsy's progress. I imagined them cringing in embarrassment to see their long-lost daughter nursing her ape-child: powdering its bottom, changing its diapers and spooning in gobs of Pablum. I used to be embarrassed too. Being of mischievous nature, I couldn't resist the temptation to place all the labels–Johnson's Baby Powder, Similac, Heinz Baby Food–in prominent positions for the camera.

Audrey Winterburn, my best friend in Victoria, was the one who took the pictures. She lived with her parents, Len and Irene, a few streets from Monterey School and I often stayed there for dinner instead of driving the Chipmunk Route home on nights when after-school meetings were late and evening meetings were early.

The Winterburns overcame their initial prejudice against Gypsy by seeing her often. Len confided to me, "I thought holding her would give me the creeps." His daughter-in-law, Kathy, who had her own newborn baby, said, "I felt sure a gibbon would be dirty." Irene was the only one who was prepared beforehand to like Gypsy. She even offered to make the gibbon diapers.

Gibbons have no need of diapers in the wild. Swinging from branch to branch as they do, they urinate and defecate whenever and wherever they feel like it. Living as a small family, there are few to mind. And being the only ape that doesn't make a nest for sleeping at night, they have no bed to keep clean.

However, as Gypsy's gymnastics increased all over her tree house, so also did the need to contain her excretions. As she had no tail it was a relatively simple matter. Some friends invited me to dinner especially so they could measure Gypsy for diapers, and later, pants. One parent told me she could make sixteen diapers from a pair of plastic shower curtains.

When Gypsy celebrated her six-month-old birthday in March, the class baked her a cake and pasted cards all over her tree house. They had done the same thing on St. Valentine's Day. To mark this stage in her development, everybody wrote poems and stories in her honour. Wendy Fischer described Gypsy as:

A squeaking, squawking little animal that twists and turns,
A shrieking, crying little animal that wiggles like a worm,
She tries to get to the top of her tree
With big brown eyes she looks at me
The classroom ape named Gypsy.
Such a difference since six months ago
You should have seen her grow!
She shot up like a bean on a stalk
As quick as she can squawk.
Soon she was stretching and grabbing
Getting into mischief and nabbing.
She wouldn't eat her baby food last week
She wanted things like carrots, fruit and meat
Things that are hard to chew on.
Soon our baby will have gone
And Gypsy will be a teenager.

Gypsy's increasing curiosity led her to explore the world around her. One of the most fascinating evenings I have spent was the night Gypsy explored the coffee table. Previously when I had placed her on the floor and sat beside her, she remained in the security of the space enclosed by my legs. On the night of her six-month-old birthday she began to move away to explore the floor space. To give her greater assurance, I lay on the floor at her level beside one of the legs of the coffee table. Having not yet learned to walk upright, she crawled, crab-like, in almost a sitting position, with her long arms stretching forward to touch the floor. She reached out and touched one leg of the table, then suddenly, as if aware of her newfound temerity, crawled back and passed her fingers lightly over my face. Within a minute she was off again to the second leg. Again she touched it, then returned to touch me. She did the same with the third and fourth legs of the coffee table, always returning to touch my face as the source of her security in between each new experience. Eventually she crawled all the way around. During the next week she widened her range even further till she had reached the edge of the carpet.

The parrots proved too disturbing for Gypsy's forays in the living room so they went to new homes where they themselves could become the centre of attention. Nothing could compete with Scarlett. She remained on her TV perch, hanging upside down and banging her play chain against the wall whenever she wanted attention. Unfortunately this was all the time. She kept up a stream of comment whenever visitors gathered in the living room to watch Gypsy.

It was fascinating to watch the gibbon teetertotter between the familiar and the unfamiliar, the old and the new, expanding her social space step by step. As the urge to explore drove her to new experiences, the fear of the new held her back to take refuge in the familiar.

As a gibbon develops from dependence to independence, it passes through three different stages.

The infant stage in gibbons is usually regarded as being from birth to two and a half years, at which point the animal stops sleeping with its mother. This is usually a couple of months before the birth of her next infant.

During the first month or six weeks in the wild when the gibbon infant is still very spider-like and naked, it rarely, if ever, leaves its mother's fur. It maintains almost continuous contact. However, even while clinging, the infant is almost continually reaching out and carrying objects to its mouth. By four months it begins to eat solid foods, and with its growing coat of hair reducing its need to cling for warmth, it moves away from the mother for short periods. But at the slightest provocation, it returns. In its second year the infant gibbon becomes more independent and begins to play with its older brothers or sisters.

During this first stage of development the infant is tolerated by both its parents. The father plays with it, grooms it and guards it when necessary. In certain captive situations he's even been seen carrying it around for as long as a week. Yet not all cases in captivity have been so fortunate. Zoo keepers think very carefully before they permit the father of a gibbon family to join the mother in raising their offspring.

In 1923 a gibbon was born in the zoo in Rangoon. In this instance the parents were not kept together and they had to copulate between the netting. When the infant was about six weeks old the mother took it over to the father's cage. He took hold of the baby's finger and was fondling it when something apparently frightened the mother who tried to pull away. The father objected and held onto the finger. In the resulting tug of war the infant's finger was pulled off and it died of septic poisoning.

In 1945 another gibbon was born at the Bronx Zoo. The family was kept together and for the first few weeks the father's conduct was exemplary. He displayed great interest in his son and touched him gently. Never once did he try to take him from his mother. Gradually, however, he grew jealous of the attention lavished on their infant by the mother and tried to get some for himself. He snuggled up to her and presented his shoulders for grooming. Still she ignored him. Finally he tried to drag the infant from her embrace and a real squabble broke out. At that point the keeper removed him to an adjacent enclosure and he had to be content with visual contact.

No animal breeds in confinement unless its needs are satisfied. The records show that gibbons give birth more readily when kept as a family in as natural a setting as possible, such as the comparative freedom of trees on a moated gibbon island with a high degree of privacy and seclusion.

The second stage of a gibbon's development lasts from two and a half years to four and a half years. This is the time when the juvenile is completely dependent of the female, is much interested in play, and is increasingly involved in competing with the adults for food.

In the third stage, between four and six years of age, the infant becomes an adolescent. Its parents, and sometimes even its younger brothers and sisters, become increasingly antagonistic. When the adolescent's urge to explore becomes greater than its urge to stay with its family, it leaves home.

Gypsy's early development was somewhat arrested by her rejection and her broken arm. However, with her many mothers and fathers, her numerous playmates and her varied environments, she grew more independent and socially mature at an earlier age than if she had been raised in the wild.

Gypsy had unlimited opportunities to indulge in play, which is all-important to a primate. Unlike her wild counterparts, she didn't have to worry about heat, cold, wet, predators, food or drink. And instead of a maximum of four brothers and sisters to play with, Gypsy had hundreds. Her play was a little inhibited in the confines of the tree house but as soon as she moved from there it was bounded only by the walls of the classroom and her own motivation.

It was the children who informed me that Gypsy, now six months old, was ready for a more challenging play area than her tree house. It was spring, not only according to the calendar but also by our own burgeoning feelings after such a long, drab winter. "Surely," said the class, "Gypsy feels the same way."

Everybody volunteered to build her a new home. We constructed a flat box four feet square, on the classroom floor, dragged over the tree stand from the Christmas play, and set to work sandpapering and tailoring a tree to fit the proportions of an infant gibbon. We cushioned the bottom of the box with beach towels, and to ensure they stayed there amid her gyrations, we attached them to the trunk of the tree. Toys were scattered liberally within the confines of the floor. Gypsy was free to stay inside the foot-high walls or climb out and crawl around the classroom. We knew that sooner or later she would be doing the latter but we hoped it was later than sooner.

Stephen Shortt marked the occasion with a poem.

Here she is, rough and tough
Biting, fighting and talking gruff,
Bouncing, boxing and swinging free
On her brand-new outside tree.

At first Gypsy lay on the towels looking up curiously, taking in everything new. Then slowly she reached out to hook the first branch, swung herself up to sit on it, then quickly descended to the floor again. Pausing for a mere moment, she swung again to the first branch and then up to the one above it. Day by day she tackled another branch till at the end of the week she was sitting at the top of the tree five feet from the floor.

As her pen was positioned beside the radiator in a centrally heated classroom, we didn't have to worry about keeping her warm. Now she was seven months old, comfortably furred, and spending her days doing incredible gymnastics. We had other problems. Her antics were too distracting to both myself and the class (there were a few other topics we had to study beside Gypsy!) so we were forced to turn our backs on her. We moved our desks to face the centre of the room rather than face Gypsy who for the first time was on the sidelines. The little imp was still within reaching distance of some of the desks and she made the most of her opportunity for contact. From the edge of her box she lunged out to touch a back, pull hair or filch a pen or pencil. I caught some of the class edging their chairs backwards to help her make contact more easily. Joan said, "She must feel neglected because she clings to the back of my desk."

Watch out, Gypsy, or the teacher'll see you!

To encourage her to explore upward rather than outward, we inserted taller, more challenging trees in the Christmas-tree stand. Daily the children decked the branches with spring flowers and baby cookies. Her favourite floral edible was daffodils. By constantly changing the tree and tying different foodstuffs to its branches, we satisfied her love of the new within the confines of the box. To our amazement she stayed inside her tree pen for almost two months.

With an abundance of toys and playmates and with a playground unrestricted by having to cling to a mother, Gypsy spent more time in play than would a wild gibbon of the same age. And she reached the main play stage earlier. At seven months she played with her hands and feet, explored and manipulated everything she could find or was given, popped it all in her mouth, spun and swung her lithe little body into innumerable fantastic gyrations, all common characteristics of infant gibbon play. But she also graduated to gibbon activities that her two- and three-year-old gibbon siblings would be doing–chasing, wrestling, slapping, clinching and biting.

As I didn't think the parents or School Board would understand that biting was a natural part of playing, we tried to curb such tendencies when Gypsy directed her teeth at her playmates at Monterey.

Ignorance and fear of the new brings out the worst in us all. I remember returning to the school after taking my class to the museum to find the nurse, stern and disapproving, waiting for me on the steps.

"Gypsy has bitten a girl in the grade two class where you left her for the day. I phoned to tell her mother. We both want to know if the gibbon has had her rabies shots. I think you had better phone Mrs. Smith right away."

My stomach took a dive. Cats and dogs have rabies shots. Skunks and raccoons and foxes have rabies shots. People don't. And Gypsy, for the intents and purposes of disease prevention, was a "people." She had shots like polio, tetanus and smallpox, just as human infants do. When faced with the unknown or the unfamiliar, people don't think. They cry rabies instead.

I was tempted to be flippant and argue. Instead I left my class and ran down to the grade two room. The victim was still drawing gibbons at her desk. To do that she couldn't be bitten too badly. I was worried more about what people made of the truth than the truth itself. I forced myself to appear casual. I glanced at the hand. Sure enough, there was a neat row of little white impressions recognizable as Gypsy's baby teeth but no blood, no breaking of the skin. I wondered what all the fuss was about.

I phoned the child's parents and tried to keep cool. The mother was sitting by the phone worrying even more and thinking things I'd rather not have expressed. "You shouldn't be allowed to have an ape in school," were the first words that burst into my eardrum. "I phoned my husband at work. He was most upset. He's been telling me all along that classes should be divided into those where parents want apes and those that don't. I also phoned my doctor. He said to check whether the animal has had its rabies shots. We'll all see what Sara looks like when she comes home at lunchtime. I tell you, though, it shouldn't be allowed."

What I still didn't understand was why all the emotional outbursts when the patient was happily crayoning apes in the classroom. Surely, if everybody felt so upset and feared there was a danger of Gypsy being rabid, they would have sent the child home immediately or to the doctor or to the hospital.

With so many visitors popping into the room to look at Sara's hand, the patient must have been wondering too. By lunchtime she went home for soup and sandwiches and the storm was over. Why do people's feelings run away with their reason? Fear, I told myself, as I shoved my stomach back in place again.

As we watched Gypsy at play we saw how her present was merging into her future. As she practised her acrobatics, she improved her speed and precision. She learned how to flee from danger with agility and judgement. As she discovered how far she could go before bringing down punishment upon herself, she learned how to mix in a group.

But we couldn't give our little gibbon the most important plaything of all. Her own brothers and sisters. She had been rejected by her own kind; and then at a time when they might have accepted her she had rejected them.

She was happy and enjoyed her stimulating environment. She displayed none of the stereotyped behaviour of bored animals in zoos–like pacing, masturbating, sucking their paws, shaking their heads, biting themselves or pulling out their hair or fur. Or the withdrawn behaviour of humans in our cities–like sitting alone on park benches, jostling in the congested streets or lying in our mental institutions.

Gypsy had passed the period for imprinting to her own species. But she interacted confidently with humans of all ages, was watched over at night by a macaw, and later she mixed with cats and dogs, not to mention bears, raccoons, cougars and coatimundis.

To prepare her for the time she might mate and raise a gibbon of her own, we did what all human mothers do. We gave her dolls to play with and we played dolls with her.

Years later zoos tried to teach gorillas in the same way: by showing pregnant gorilla mothers films of proper maternal behaviour, by giving them surrogate infants to handle, and by teaching them voice commands like "Pick up the baby, Dolly", "Be nice to the baby, Dolly", "Turn the baby around, Dolly."

At a time when little was known of any ape in the wild, at a time when they rarely lived long or bred in captivity, we raised Gypsy to be as well adjusted a gibbon as it was possible to be.

6
Sometimes a Gibbon IS a Monkey

Gypsy, the gibbon, come out,
I haven't got time to play
It's supper time and I'm hungry
Gypsy, the gibbon, come out.

Gypsy, the gibbon, eat your food,
The bell is soon to ring
Children rushing in and out
Whatever am I to do?

Gypsy, the gibbon, come off that,
Soon you'll be flat on your face
Then you'll be crying and screaming away
Yet I warned you to come down off that.

Gypsy, an angel?
Ha!

(Kathy Warren)

The children echoed what many of their parents must have said when their own offspring were babies: "Oh, isn't she an angel when she's sleeping!" and "Isn't she a monkey!" when she wasn't.

Seven months old now and out of the incubator, Gypsy rarely slept in school. But when she did, we enjoyed watching her expressions. She was restless even in sleep. Her eyes rolled around under almost closed lids. She breathed erratically and emitted little squeaks. We wondered if she dreamed. And if so, of what? Of a Thailand jungle? a concrete jungle? or a blackboard jungle?

While riding to school in the car she was no longer content to sit quietly in the cardboard box (the Styrofoam one had long been chewed to oblivion). She would climb to my lap, sit on my shoulders, swing on the steering wheel by her arms or else by my driving arm, and at the same time pump the indicator light with her feet. Luckily, no policeman stopped to ask why I weaved.

One day I was horrified to find her out of my sight. I swerved over to the side of the road and found her casually stretched on the floor between the seat and the door, investigating specks of dust on the carpet. After that, I had to keep the car windows closed or she would have crawled out to catch the windshield wipers.

Gas jockeys loved her. But the attention they gave her was sometimes annoying. With everyone's eyes on Gypsy, gas spilled out of the tank and gas caps were left behind at the pumps.

At school she continued to devise new games for herself, compromises between the now-familiar tree house playpen and the new world outside its borders. From the lowest branches she still swung out to bounce off the backs of the nearest pupils, tug at their braids or rip off their rulers. But now that she had progressed to the topmost branches which were level with the window ledge, she discovered to her delight that she could pull at the cords hanging from the window blind. The first time they went crashing noisily to the ceiling she sent herself crashing to the floor to cling to the bath towels for comfort. Within seconds she was up the tree to try it again. So we tied the cords in knots and kept the blinds raised and out of reach.

Soon we had to clear the window ledge completely. With my classroom so crammed with learning materials and the creative efforts of the children, I had to keep my files, individual reading cards, library books and the lost-and-found box on the shelf formed by the top of the radiator and the window ledge. But this became Gypsy's domain, at least the part she could reach with one arm or leg. She kept the other half of her limbs in contact with the tree for security.

It was clear that Gypsy would soon abandon her playpen altogether. In readiness for the chaos I expected that stage to bring, I began to lecture the class more and more on tidiness–especially when the janitor complained to the principal that the monitors had been tossing papers and cans left over from feeding and cleaning Gypsy, not in the overflowing class garbage can but in the nurse's one in the medical room.

"Looking after Gypsy is a privilege," I lectured grimly. "You must be specially clean and tidy. We must be very careful we don't cause anybody but ourselves extra work in the school. Or else Gypsy'll be asked to leave."

Remembering how the school year had started, I refrained from mentioning that both of us might be asked to leave.

The Easter holidays probably came just in time. It provided a respite for everyone but me. The term ended as usual with my next bout of laryngitis and my mother-in-law advising rest, a tonic and iron pills. Gypsy also had a change-of-season wheeze. This time I didn't send her away, as I hoped by now she was more resistant to infection. And with her home range expanding at its current rate, I didn't like imposing on the babysitters of her incubator days. Instead we sweated out our colds together.

As her favourite position was being wrapped around my neck anyway, I boiled up a steaming solution of Friar's balsam, enclosed the two of us in towels and lowered our heads over the bubbling pot. Gypsy didn't like it. Neither did I. But I kept telling her, as my parents had told me, that we had to put up with it because it was good for us. I don't know if that did it or not but in a few days we had both recovered and were ready to cope with the vacation.

Spring break was the time for putting Scarlett out on her totem-pole perch by the front door where she could amuse visitors to the beach; it was the time for doing a deep clean in the rooms, sprucing up the compound and, this particular year, painting the house.

It was also the time when we were inundated by both people and animals. In this season of otters digging nurseries under Oak Bay basements, raccoons scratching tunnels into Brentwood cellars, nestlings toppling from trees and bear cubs being abandoned in the bush, more and more people came seeking help. Others just wanted a destination for a sunny weekend drive. Visitors either brought their own animals or wanted to see ours. On Easter Sunday an article I had written for the local paper attracted over a hundred sightseers. They would not be turned away by my pleading that we were not open to the general public on a regular basis.

Despite her professional prowess as an entertainer, Scarlett couldn't keep everybody amused. So I found myself with paint pot in hand going up and down the ladder like a yoyo. One lady was indignant that the animals could not be seen behind the ten-foot-high compound fence. She stood underneath me at the bottom of the ladder and called out, "And where *are* the animals?" Just then one of the animals nearly upturned a pot of Dayglo orange paint on her head. The window to the living room swung open, nearly knocking me from the top of the ladder, and with a *whoop!* Gypsy swooped onto my head and encircled my neck with a vice. Hearing my voice so close to her perch in the inside tree house, she had for the first time pushed up the sliding door at the back of the box, leapt through the curtains and shoved herself through the window. The paint pot jiggled, the ladder wobbled but everything stayed in place. Except the lady.

The sight of a black hairy ape flying through the skies as if on command must have made her wonder what else was lurking in the living room beyond the paint pot. Scarlett wished her well with a cheery "Good-bye!"

Once she had me in her clutches Gypsy was happy. So between visitors I continued to paint the house–and her! As soon as she was confident enough to release one arm from her grip on my tonsils, she used it to follow the paint brush to the wall. Then, curious to see where it came from, followed it back to the paint pot.

"Get your fingers out of there, Gypsy," I admonished. One of my hands held the pot, the other the brush, and my vision was blurred by her agile body. She couldn't have liked what she found. She shook her head and hand in her typical gesture of distaste, flinging orange paint over both of us. I wiped her face and fingers, ignored my own and tried to continue painting. After that I only had to cope with getting the paint brush to the paint pot in between her flying gyrations as she swung under and over the rungs. Finally she climbed the ladder above my head and was content to keep contact with her toes.

Hilary arrived the next day to help me cope with the chores and the holiday visitors. Gypsy went with me to the bus station to pick her up. With the streets blocked due to an accident, the parking lot beside the terminal full, and the sidewalk congested, I slipped into a taxi rank space and hoped not to be noticed. Hilary was sure to be right there at the door and we could make a quick getaway. No sooner had I opened the car door than Gypsy slipped onto my shoulders and with no time to brush her back into the car, I ran into the bus station.

With a black ape clinging to my hair and screaming in excitement at the noise of the traffic and the people, I was sure to attract attention. Fortunately, Hilary was the first to notice. She grabbed her suitcase and dashed after me to the car. The sight and sound of Gypsy hanging from my neck silenced an irate taxi driver, and I took advantage of his astonishment to accelerate from his stand in haste.

Gypsy responded to the excitement in the same way that her wild counterparts would when they suddenly came across another family in the forest. She made a lot of noise. Gibbons make more noise than probably any other animals in the entire animal kingdom. When one hears their chorus of ascending whoops one cannot remain a passive listener. One must join in, drown them out, shut them up, or leave them vocally victorious.

Hilary arrived at a time when Gypsy was widening her social circles. Upon our return to the house for the holidays, the gibbon obviously missed the attention lavished on her during the school day despite her peripheral positioning in the classroom. All the rooms at Island View abutted the living room but if I ever disappeared from her sight to work in another room, her excited *oo-oo-oo*'s rose to a screaming pitch. When I took her out of her tree house at meal times, she attempted to lengthen the feeding process by shaking her head, pushing away the food and making little *uh-uh-uh* noises as she flung her limbs about. I tried to pop in a spoonful whenever she opened her mouth but feared to make her choke. And as eating in such excitement usually caused her to regurgitate, I had to think of a compromise.

I had to find a way to get my work done yet give her attention to compensate for the companionship she would have enjoyed in the classroom. Gypsy had always accompanied me when I drove. Now she went everywhere I walked. Everything I did from then on was complicated by a gibbon hanging from my neck. It would have been easier if she had decided to go piggyback but gibbons in the wild don't do that, nor do gibbons in captivity. So Gypsy clung to my chin.

One morning a couple of days after she had interrupted my painting schedule, I awoke to find her opening the sliding door at the back of her tree house. Oblivious to the weight of the door on her back and still with one hand cautiously clutching a familiar tree hold inside the incubator, she squeezed through the opening and stretched over to the curtains. After a moment of indecision and some soft *oo-oo-oo*-ing, she finally relinquished the branch and swung with both hands onto the drapes. Bunching the material between her fingers and toes, she inched her way over the unfamiliar expanse to the curtain rod at the top. Scarlett with her beak buried in her feathers saw her coming. Startled to consciousness, she edged over to the far side of her perch and unthinkingly squawked, "Good-bye!" obviously hoping the gibbon would go away. If she had remembered her manners she would have said politely in her best Oxford accent, "Hello." I couldn't resist laughing as Gypsy, startled, leapt to my neck as soon as I got within reach.

Next morning while I was outside thawing herring for Sam and the seabirds, Hilary woke to find Gypsy out of the incubator, clinging on all fours to the curtain and screaming at the unfamiliar world. This time Scarlett was bending over her, asking all the right questions: "Hello and how are you? I'm a parrot. You're a bad bird." But now Gypsy probably wished Scarlett would say "Good-bye" instead. Hilary rescued her and they had a leisurely breakfast of milk, apple and banana. As soon as I entered and Gypsy heard my voice, she screamed again, jumped off Hilary's lap and clung to my leg till I lifted her to my neck.

That same day I was at the stove with Gypsy as usual around my neck and Hilary standing close by, helping me cut up vegetables. Suddenly Gypsy reached over and in an unexpected and apparently unthinking outburst, bit savagely through Hilary's sweater to her skin. Her teeth weren't strong enough to bring blood but I was astonished by this first demonstration of a vicious unprovoked attack.

Perhaps because of Hilary's very closeness to me Gypsy attacked through jealousy. Perhaps it was another temper tantrum marking her displeasure at lessening attention. Perhaps it was the turmoil she felt at a time when she was exploring new places. It was Gypsy's custom when adding to her territory to seek additional security. At school any number of substitutes were accepted. At home it seemed she only wanted me, except for the few occasions I was not in sight and somebody else would do. Despite these rationalizations, I still suspected there was something different in her attitude towards Hilary.

After Hilary left I made a special effort to expand Gypsy's social contacts. My diary of the time reads caustically but without comment: "High jinks at the Winterburns, screams at the Dawsons, fun and frolic at the Pattersons." And then with a little more impact: "Gypsy crawled around the floor of the municipal hall to the amusement of the receptionist", "The women of the women's page of the *Daily Times* loved Gypsy's visit", and "I changed Gypsy in the ladies' fitting room of Sweet 16." And: "Gypsy helped me kiss the Bible when I swore the Oath of Allegiance."

At that time in my animal career I still made space in a busy schedule for the theatre which until my marriage had been my first love. Gypsy always came with me when Audrey and I went to a play, a concert or a lecture at the museum. By nightfall and in a darkened theatre, Gypsy would sleep in my arms or in a shopping bag at my feet. Hidden or at least disguised by my coat, few saw her until the lights came on and by that time I had checked in at the washroom. Anybody in the audience who did see her was usually enamoured.

Except one night at the McPherson Playhouse. Having come in after the performance had started and not wishing to disturb anybody by looking for a seat in the dark, I stood up at the railing behind the last row of seats with Gypsy asleep in a blanket in my arms.

We made no sound, but a middle-aged lady in the seat directly in front happened to turn around. She must have been eagle-eyed to see so clearly in the dark. With one horrified look, first at Gypsy and then at me, she immediately got up and moved two rows down. At intermission, instead of beating a retreat to the washroom as was my pattern, I stayed in the auditorium, intending to introduce myself and allay her fears. I expected her to croon *ah* and *oo*, like everybody else did. But I had no chance. She saw me coming. I was about to offer her a cradled ape when she shivered, showed me a very cold shoulder and said between her teeth, "How did the management let you in with that animal? It shouldn't be allowed." I was stunned by her attitude. She had gone before I had a chance to tell her that it was my film she was watching. For that lady the reality of the screen was close enough. She didn't need to meet any of the actors in person.

Although our visits were short and infrequent, one of Gypsy's favourite places was the doctor's office and waiting room. If I met the doctor on the street or at the theatre, he always asked after his favourite patient and I always kept him informed of Miss McGibbon's latest measurements. But though he watched over her progress and was interested in her condition, he had no need to diagnose any complaint or prescribe medication as Gypsy, apart from sharing a few of my colds, had stayed healthy despite the well-known delicacy of her species in captivity.

Miss Gypsy McGibbon went to the doctor to entertain the patients and watch over me. One day her solicitude caused complications.

The doctor greeted me at the door with his usual joke, "Look what happens, Lyn, when you go off your birth control pills." He winked and gave Gypsy a chuck under the chin. "What is it today? A half-yearly examination? All right, you'll find the room empty. I'll see you in a few moments."

Miss McGibbon left me alone long enough to get undressed and to pile my clothes on the chair in the corner. But as soon as I lay down naked on the examining table, a loud *oo-oo-oo* sounded and she made a flying leap from the floor, bounced off the sink and spreadeagled herself on my neck. When the doctor entered, she was lying stretched between my breasts, with her thumb in her mouth. Although the parts of my anatomy that interested the doctor didn't conflict with the parts that Gypsy was sprawled across, the gibbon was determined not to share. As soon as he bent over me she let out a deafening *whoop* that skyrocketed to the top of the scale and screamed into the sound barrier. Up stood the doctor and in rushed the nurse to stop what she probably guessed was an unpremeditated attack on a patient by a mad doctor. It was what Miss McGibbon was doing to my lungs and stomach as she jumped violently up and down on top of them that worried me. Not that we weren't all in danger of cracked eardrums as she did her giddy dance of excitement, shook her head and her arms, wildly smacked her lips, and bounced me like a trampoline.

"Let me hold her. I've been calming babies all my life," crooned the doctor when he recovered his composure. Before I could warn him to the contrary, he had caught Gypsy in the middle of one of her gyrations and was walking around the examination room rocking her back and forth in his arms. I fully expected her to bite him as she had Hilary but perhaps because of the space that separated us, she just wound down to a moan and noisily sucked her thumb.

"There, there, now don't you worry. I'm not going to hurt your mother," he said soothingly to the ape. "I'm going to let you sit on my desk in my office while I finish, I mean, start, making sure she is all right."

I could hear him comforting his favourite patient behind the closed door. "Now you stay there, Miss McGibbon, and check my files while I check your mother." A few seconds later he had carefully shut the intervening door and was back with his stethoscope bending over my body.

Another terrible scream vibrated through the building. Then a crash.

"Oh, no, she's broken something," I wailed from my helpless position on the examining table. "I should never have let her out of my sight."

"Well, well, well, that's the first baby that has not fallen for my fatal charm," the doctor exclaimed in wonderment. "We'll have to see what happened."

He was back within moments. "The nurse is distracting her attention in the waiting room with the other patients," he said cheerfully. "I thought that was best. They might be thinking the worst of me. And there's not much damage. We'll send you the bill. Just a vase."

"What kind of vase?" I wailed again.

"Oh, nothing to worry about. Ming dynasty, I think it was."

7
Black Bears and Coatimundis

I think it was Thoreau who once said, "If thine enemy offend thee, give him a bear cub."

I hope Dr. Conway, an animal lover from Port Hardy, didn't have that in mind when he sent me Skookum and Cultus, two orphaned black bear cubs. It was fortunate they arrived in the holidays as black bears in the house demand more attention than a troupe of gibbons and a flock of macaws.

Bears are the most rambunctious creatures I know. Anything up must come down and anything in must come out, more to satisfy their sense of destruction than to assuage their curiosity. Bears want to test their muscles more than their brains. When the object they are after is wrecked, they lose complete interest in it. And when you try to remonstrate, they sit back on their wobbly feet, peer at you through little pigs' eyes and look astonished, even stupid.

Despite their furry, roly-poly, humped little figures, at feeding time they look and sound more like human offspring than any other animal I know. Human foster parents rarely resist their insistent baby cries, their licking and slurping and sucking, their rolling over on fat tummies to be rubbed and caressed.

Now look here, bears are not all that bad!

For me the attraction doesn't last long. They have gargantuan appetites and it doesn't matter how much you feed them, they will always want more. Immediately after a meal they will continue to cry, exactly like a human baby, as if they hadn't been fed for days, then climb your legs in a series of hugs, squalling incessantly for both food and cuddling. But they are just as likely to bite the hand that feeds them, even in between slurps of their long curled tongues.

But disciplining bears is difficult. They seem to grow quicker, get stronger faster, and if you slap them, more often than not they will slap you right back. Bears are belligerent. In raising them you learn the truth of the old clichés, "hungry as a bear", "angry as a bear", "strong as a bear," but I've yet to learn the reality of "cuddly as a bear." The only cuddly bear I know is my stuffed koala.

I must be fair. Not all bears are the same. Black bears come in various colours–not only shades of black but brown and bluish and cinnamon to white. Skookum and Cultus were easy to tell apart. Although they both belonged to the black bear species, Skookum was a golden reddish-brown and Cultus the more common black with a white diamond on his chest.

Their personalities were as different as their pelages. To put it simply, Skookum was good, hence her name; and Cultus was bad. Though the same age, probably three months old, and almost the same size, four pounds and about 18 inches high when they stretched up for their bottles, Cultus was the more belligerent and drank two, three and four times more than his sister. When he wasn't sucking, he was biting, fortunately on Skookum's ear, not mine.

Although I had to feed them every four hours at night as well as during the day, they were young enough at the time of their arrival to mostly sleep in between. After each meal they stood up, shook themselves of the milk that clothed their fur as well as everything else, burrowed down among the blankets in their communal box and sucked themselves to sleep.

Others thought they were adorable. The class came out from Oak Bay on their bikes to see Gypsy and stayed to see the bears. The Winterburns baby-sat them while I was downtown doing errands. And Susan Sandeman-Allen, the 17-year-old who sometimes came to help me with the chores, took Skookum home to her family for a week to compensate for the poor treatment she was getting from Cultus. That holiday didn't last long as Susan's mother phoned in a few days to say that she had been "threatened" by an inspector at the Fish and Wildlife Branch for harbouring a bear without a permit even though the bear was held in captivity under a legal permit by me. One of the neighbours had asked to come over and see the bear cub, had enjoyed her visit, then returned home to phone in a complaint. Poor Skookum, the good little bear, had to be brought home.

Nevertheless, Sue and her family of animal lovers rallied around to help when we took both bears down to the McPherson Playhouse to present our film *Coast Safari*. For most of that winter David, his parents and Gordon Cooper had been away on tour with our first wildlife film, up north in British Columbia, out east in the Okanagan and down south in California. Now in the spring they had returned in readiness for the climax of the season in Victoria and Vancouver.

Believing that personal contact is an important key to persuade people to develop a love for wildlife and a desire to protect its habitat, we brought as many animals as possible to the theatre, commensurate with logistics and the needs of the animals.

Although the film audiences to date had been enthusiastic, their numbers had not been large enough to support the substantial expenses necessitated in playing the big cities. Behind us was the winter's experience of playing in country school gymnasiums, giving free lectures in the classrooms and naming the film *The Wild Lonely Coast*. Now we knew that our images of education and conservation spelled death for large and sophisticated audiences. People didn't go to the theatre to be educated, they went to be entertained. It didn't matter that they were entertained when they got there. What did matter was saying the right things to get them there in the first place.

Of course, if you have enough money to whitewash the air waves you don't have to say anything except the name of the film, the date and the place of its presentation, but you must say it over and over again. When Gypsy and I went to Seattle to present our film under the sponsorship of the school P.T.A.'s, we were annihilated financially by a competitive film called *Cougar Country.* "Cougar Country . . . Cougar Country . . ." hit us as soon as we got off the plane, when I turned on the radio, plugged in the television or opened the newspapers. And that kind of advertising costs more money than a teaching salary can provide.

Yet the name of the film is important. All winter David, his parents and Gordon had been trying to attract people to *The Wild Lonely Coast*. Killing must have been in fashion that year as audiences were more interested in films titled *Alaskan Safari* and *Northern Safari*.

The first featured endless sequences of the same shots–hunter shooting the biggest trophy, closeup of bullet and gun and bloody wound, smile on face of hero as he sits astride his victim, other hunters congratulating him on his prowess before he flies south to hang his trophy on his office wall.

The second film featured a little black-and-white terrier labelled "the unparalleled wonder of the canine world" that yapped, menaced, bit and killed every animal the cameraman could find on a trip across Australia.

I like eating meat so I can't say I'm against killing if it is for food. But with the increasing sophistication of guns, telescopes and radar as well as helicopters, snowmobiles, bush planes and factory ships, hunting animals by such modern methods doesn't seem very sporting or even something to be proud of, as it once was for our evolution. I have always found it more difficult and time-consuming to get close to a wild animal with a camera than with a gun.

However interesting the adventures of a man and woman might be while studying live animals on a wild, lonely coast, it required a safari to see them in order to attract audiences to the film version. *The Wild Lonely Coast* became *Coast Safari* and in Victoria and Vancouver we were overwhelmed by its popularity. It had been with great trepidation that we rented the McPherson Playhouse and the Queen Elizabeth Theatre for three nights. And it was a great relief when so many enthusiastic people showed up that we eventually ran the film for three weeks.

Each showing was a friendly family affair. Before the film began, during intermission and after it ended, I mingled with the audience to chat about the animals. In Victoria's more intimate setting we spilled out onto the plaza. Susan had Skookum on a leash, her brother Glen had Cultus, three of my class had the tortoises and I had Gypsy. The gibbon stayed awake for matinées but was usually asleep on my neck for evening performances. Scarlett circulated too on the shoulder of Sue's other brother, Scott. The Sandeman-Allen family were Scarlett's favourite visitors and they sometimes took her home for weekends. The macaw rode their shoulders while they seeded their garden, attended to their apiaries and sat in Church. The relationship was a little unsteady when Scarlett blasphemed but it revived in time for the showings at the McPherson.

With our spirits considerably heightened by the heartwarming response at home and the animals obviously thriving on the attention they were getting, we all looked forward to showing the film in Vancouver. First David went with Skookum and Cultus to arrange the promotion. Gypsy and I, with chores to do and lessons to prepare, followed a couple of days later. Susan and her family stayed at Island View Beach to look after the animals and open the coffeeshop there on weekends.

Although Gypsy had often travelled with me in the car, this was to be her first ferry ride. It was hot, an unusual spring temperature of 80°F, but I was forced to wear my blue Eskimo parka because it had a voluminous hood that completely enveloped an eight-month-old gibbon. Gypsy's diminutive, human-like face was scarcely noticeable; the same could not be said for my bulging bag of baby paraphernalia.

Confident on the outside though feeling a little nervous on the inside, I bought my ticket, proceeded along the gangway, boarded the ferry and climbed the stairs to the top-floor lounge. Any minute I expected a uniformed official to tap me on the parka hood and repeat the announcement, "We regret that animals are not allowed above the car deck." If that happened Gypsy and I would be relegated to standing up for two hours squeezed between parked cars on the wet, windy bottom deck.

Fortunately, a friendly, attractive Australian helped me in my subterfuge and we spent the rest of the ride together in an isolated corner of the lounge. Gypsy was fascinated by his beard. The only drawback to the journey was when the gibbon needed to go to the bathroom. As the ferry was crowded, especially the limited toilet facilities, I had to change her in one of the toilet cubicles. I hung her on the door in my shopping bag, sneaked out to get more paper towels, then sat on the toilet seat with her on my knee to freshly line her plastic pants. I bundled her back in my parka hood to rejoin our bearded friend who offered me a lift downtown when we reached the mainland.

Travelling was not so easy in Vancouver. Accustomed to taking Gypsy on city streets, I completely forgot that she was clinging to my neck when I boarded a bus at Georgia and Granville. With my mind on other things, I handed my token to the driver.

"No animals aboard my bus, lady. Off you get."

He didn't look like the type for me to make some jest about colour discrimination for babies. So I protested instead about the number of times I had travelled by bus previously in the same circumstances and nobody then had questioned my companions. The only ones amused were the chuckling and sympathetic strap hangers standing nearby.

I got off and boarded the next bus–this time with no comment except from the girl who sat beside me. To my great astonishment, she had once raised a gibbon herself which she claimed was toilet trained. Unfortunately she had to alight before she could tell me the details and I don't remember her name.

Gypsy, despite being jostled by strangers on crowded ferries and sitting beside them on jammed buses, remained surprisingly oblivious to the unaccustomed sights and sounds of the big city. We got off the bus at Point Grey and walked to Hilary's apartment. At the top of the stairs I tugged at her distinctively embroidered bell pull and she opened the door.

"Lyn! I didn't expect you till later," she cried loudly, flinging her arms around me in an enthusiastic welcome.

"Hilary!" I shrieked as I embraced her, forgetting Gypsy who was squashed between us.

"Ouch! She got me," I heard as my hostess retreated into the living room. Once again poor Hilary had been bitten. This time we blamed ourselves, not Gypsy, who reacted automatically to being crushed, even smothered, between two armpits.

It must have been a sudden impulse brought about by a moment's fear because Gypsy spent the rest of the evening scampering unconcernedly across the floor of the living room. She was now almost walking upright though her long fingers still swished along the ground on each side to help her balance. Soon she was lightheartedly running from one end of the apartment to the other, pushing everything off tables and onto the floor, climbing up on the trilight to reach for my wig which perched atop the lamp shade, pulling down a piece of the rubber plant and trying to poke the fish in the aquarium without getting her fingers wet.

It is well known that gibbons dislike or fear water. Some scientists have demonstrated this by placing gibbons in tanks containing ten feet of water. The animals struggled helplessly like children do when learning to swim. Zoo keepers, knowing gibbons will not leave the land even if freedom is only two feet away, take advantage of this aversion to water by isolating them on monkey islands. Lee Crandall of the Bronx Zoo described how one gibbon missed his grip on an overhanging branch and drowned in water not more than two feet deep and three feet from the island. While it floundered helplessly, its mother watched from the shore but made no effort to assist it.

For the first few weeks of her life Gypsy had been surprisingly tolerant of being bathed in a basin, then suddenly for no accountable reason her behaviour changed and she screamed as soon as we tried to immerse her. As she kicked soapy suds at us and squirmed out of our hands when we tried to restrain her, we always finished up getting wetter than she did. When we moved from the mixing bowl to the sink in the girls' washroom, bathing became more of "a cat's lick and a promise" of pat and dab than clean and rinse.

Gypsy hated water and would shake her fingers vigorously if she got any on her. Yet as soon as I left her with Hilary and the others in the living room and went to the bathroom for a shower, she ran after me. As I pulled my clothes above my head she dived for my legs and stayed there. She was still clinging when I stepped into the bath and turned on the taps. I was sure she would relinquish her grip and climb out as soon as the warm water hit her, but amazingly she just stayed there, a tight little snake, her arms wrapped almost twice around my leg. She even pursed her lips and sucked up the water that was drenching her. She was still there when I stepped out of the tub. It was even more incredible that she let Hilary take her back to the living room and dry her in a towel while I continued dressing in the bathroom. If she feared Hilary or feared remaining in a strange room without me to the extent that she could overcome her natural fear of water and follow me to the shower, then why did she let Hilary take her back to the strange room without me?

Later that night David who had been taping a television show arrived with Skookum and Cultus and Scarlett. The macaw greeted us with a husky "Hello" and was rather unceremoniously put out of mischief's way on a top shelf denuded of its breakables. The bears climbed out of their box onto the coffee table and sat there wobbling while they took in their new situation. Gypsy couldn't reach Scarlett's tail which hung down the wall just above her head but she darted over to greet the bears.

Always thinking first of her protection, I had allowed Gypsy few opportunities to play physically with other animals. Not yet having free rein of the house at Island View but spending most of her time in the tree house or attached to my neck or leg while I was doing chores, she had been easily separated from the bears and kept her distance from the curious macaw except when Scarlett's back was turned and she had a chance at her tail. But as we were all to spend the next few days crammed in Hilary's living room, it was a good time to test our co-existence.

For a moment I worried. As soon as the bears caught sight of the bouncing gibbon they raised themselves to their full height of a couple of feet and batted her. Gypsy did the same. She stood upright, swung her long arms above her head as if to make herself bigger, then batted them back.

It was the funniest sight I had seen in a long time. The trio stood jumping up and down facing each other like boxers. Especially Gypsy who dashed nimbly in to fling her fingers at their heads then danced back out of the ring again. Poor Skookum couldn't keep up with her hit-and-run tactics. The bears, being more stolid individuals, just grasped at air in her wake. They declared her the winner and rolled over together in a clump in the corner.

To our surprise, Gypsy walked over to the cubs, wound her long arms around first one, then the other, and gave them the gibbon kiss. It was the same embrace that she gave me. It was the greeting behaviour that Dr. Carpenter noted in gibbons in the wild. When two friendly gibbons approach each other they draw back the corners of their mouths into a smile, bare their teeth and sometimes protrude their tongues. If they meet while walking, they embrace with both arms. If swinging, they clasp each other with one arm but both feet. Though I never heard Gypsy give it to an animal, Dr. Carpenter reported hearing wild gibbons accompanying their greeting with a little cry like a faint squeal.

Gypsy, though taller than the bears when she stood up, was about the same weight so all three animals were pretty well evenly matched for size. However, the bears, though stronger, were no match for her quickness and agility. As soon as her polite introduction was over, she chased them up and down the living room, slapped their backs then darted away before they really knew who had flashed by. She paused at the door to leap up the wall and tried to grab Scarlett's long scraggly tail feathers. Highly indignant, the macaw squawked "Stop that!" a response that never failed to put us into hysterics.

"Shush!" whispered Hilary through her laughter. "Remember the neighbours."

They would have to be new neighbours as few weeks went by when she wasn't hosting one or more of our critters. Her apartment was full of animal delights, soft foamy cushions, whole walls of shelves, a terrarium of frogs and salamanders, chairs like hammocks, artistically arranged twigs, a mini jungle of hanging and potted plants, even a cow skull in front of the fireplace.

Gypsy learnt her way around very quickly but when she turned her attention from teasing the bears to testing the carvings on the shelves, we deemed it time for bed. The bears were stowed in the laundry basket, Scarlett was happy to stay on the door and Gypsy lay quietly beside me in a box by the chesterfield.

Gypsy now showed no signs of hostility towards Hilary so the two of us agreed to take her in the car the following day while we put up posters to advertise the film. David with the bears and the macaw drove off in his car to finalize arrangements for the big show.

Everything went well all morning. We visited my familiar haunts from the days when I lived in Vancouver, called on some friends from a former class who offered to help with placing posters or looking after animals during the performance, and enjoyed talking to the merchants who used to meet cougars or a seal when I came to shop. Now they met Gypsy and were entranced as she curled her arms around their necks and gently kissed them in greeting.

Back in the car we settled down to plan our next place of attack, and to our complete astonishment after the calm uneventful morning, it happened to be Hilary's arm. Without any warning at all and unprovoked by anything we could understand, Gypsy suddenly leaned over the gear shift and dug her canines into Hilary's driving arm.

Her teeth still belonged to her baby set and didn't do much damage but the bite hurt. I was nonplussed. She couldn't have been afraid of strangers. She was sitting on my lap in the passenger's seat and had been perfectly amenable all morning, as she always was with everybody but Hilary. Was she jealous? Did she remember being squashed between us the night before and was revengeful? Was she confused by a different person in the driving seat? Or was she trying to protect me from imagined dangers?

Whatever the motivation, I kept a tight hold on her while we drove to the next shopping centre. Once there, she enthralled everybody again by giving them gentle kisses.

It does not appear that gibbons are aggressive in the wild. Vocal battles or displacement activities, like jumping up and down, leaping between trees, and shaking branches, tend to substitute for actual physical contact when groups meet at their territorial boundaries.

In captivity social groupings are enforced, and where many unrelated gibbons are confined together, fights occur regularly as frustrations are greater.

Certainly Gypsy was forced to meet people every day of

her life but she actively sought attention and was always affectionate. Even in harassing circumstances, like seeing the doctor examine me intimately or being rushed through a noisy bus station, she had been excited but she had never bitten anybody in spite. Nobody but Hilary who remained one of her staunchest supporters!

The next day we took a holiday from film promotion and drove out the Fraser Valley to the University of British Columbia Research Forest at Loon Lake. It was a quiet, scenic setting for filming the bears. While Scarlett skittered around the roof of the car, swung on the rear-vision mirror and yelled "Yoo hoo," Skookum and Cultus made the most of the forest. Their figures had grown with their appetites, their wobbling walk was now a flat-footed gallop and they were soon humping up trees with their strong non-retractile claws. But as soon as they got to the top and we were hidden from sight, they cried and came down. Their descent was a little slower than their ascent.

Gypsy enjoyed the unaccustomed space. She ran everywhere, her long arms skimming the ground for balance, and was only limited by her distance from me. When I wandered too far away for her sense of security, she made her little plaintive *oo-oo-oo*'s till she had caught up. She landed on my back when I leaned over a pond to pick out tadpoles, and sat on my camera when I tried to photograph wildflowers.

The bears were more independent. They waded into the water while chasing butterflies, lost their footing in the mud and discovered they could swim. When they hauled out on a log that was floating handily by, they almost learned to birl. What a difference when they clambered, dripping, up the bank! Their little rotund figures from the pages of *Winnie-the-Pooh* had shrunk under the water so they looked like skinny beanstalks with oversized heads. A good shake, a cuddle from the gathering knot of backpackers standing by, a few more pulls at the bottle, and Skookum and Cultus went exploring for more.

When we returned to Hilary's apartment, there was a message on the table. "There are two more animals to pick up at the airport. (signed) Hilary."

"I thought life was getting too easy for you with just a gibbon and two bear cubs," said David with a grin. "The Calgary Zoo phoned a few days ago and asked if we wanted to look after a couple of coatimundis. I told them you needed the experience."

"Looking after what?" I asked with my usual groan at another arrival.

"Coatimundis," he replied as he jiggled his car keys. "A South American kind of raccoon."

And so Pixie and Pete were the next guests to burrow their way into Hilary's apartment and into our hearts. She offered them accommodation in a cardboard tunnel in a vacant terrarium and being very tiny and only three weeks old, they happily buried themselves in the folds of a bath towel till feeding time.

The word "coatimundi" has a Latin derivation which means "nosey one," an apt term because one of the coatis' chief characteristics is its long, saucy upturned snout. The other is its long, slender ringed tail that helps it balance while climbing and often shows how it is feeling. A coati looks something like an elongated pig in front and a raccoon behind except that it carries its ringed tail erect.

Although its appetite has been described as "virtually unappeasable," or less politely, "like a travelling garbage can," there was little to indicate that feature in the days Pixie and Pete were on the bottle. In contrast to Skookum and Cultus who downed one 16-ounce bottle in a few noisy seconds then cried for more, the new arrivals fussed so much in getting started on an ounce per feed that I ended up holding their long snouts over the bottle and forcing them to eat. They soon made up for their early diffidence and like the bears, seemed to sprout before my eyes.

Gypsy greeted them with the delicacy their tiny size de-

Gypsy greeted all newcomers with a kiss, though Pixie is probably taking a sniff.

served. This was the first time she had introduced herself to so small a creature and it was exquisite to watch. She reached out tenderly and enclosed each of them completely with her body, not only her arms but her feet as well. She was like a mother cuddling a child or a child a toy. Pixie and Pete didn't have a chance to get away even if they had wanted to. She had them clasped in a ball within all four limbs.

Next morning we all went to Pete's Place, a popular TV show produced in Burnaby at British Columbia Television where Hilary worked. Scarlett. Gypsy. Skookum and Cultus. Pixie and Pete. As usual it was chaotic and enjoyed by everybody but me.

We should have taught Scarlett to say "Welcome to Hancocks' Wildlife Adventure Tours" and then she might have given point to the program. But as one of those words often reminded her of one of her embarrassing obscenities, I never took the chance. And so the program became a scrum, with me trying to stop Gypsy from enfolding the bears who were trying to climb all over the interviewer who was trying to show Pixie and Pete to the cameras. A closeup of the coatimundis showed elbows, shoulders and knees, claws and paws and diapers, instead of a pert twitching nose and a long ringed tail.

For sanity's sake I should have shown one animal at a time but as they all lived together at home and in the field, I could see no reason to confine them in cages just for television. The production staff must have been in a panic too, as the film clip I was supposed to be narrating while the animals provided the visual background got cut in half and the slides came out upside down. Afterwards Hilary said the action was fantastic and that was all that mattered. Fortunately, Pete (the professional one) remembered his job and plugged the film while the rest of us disappeared out of view in a merry chase under the chairs and coffee table.

Next evening was opening night so we took the animals to Stanley Park for a workout before the show. It was as riotous outside as it was inside the studio. Several groups of picnickers rushed over to watch when Skookum and Cultus climbed trees. I kept the crowd at a distance, so the bears would come down when their burst of energy had dissipated itself. The coatis seemed to have grown overnight, at least in behaviour, as to my surprise they started to chase Gypsy who fled to my neck.

The children thought it great fun and wanted to watch all afternoon. With a glance at my watch I realized that if the bears stayed in the tree I would have to call a fire truck to get them down in time for the performance. Reluctantly the audience returned to their picnic tables. I collected Gypsy and Pixie and Pete and walked away from the trees. My ruse worked and the two little bears soon scurried after me.

Just in time. We all piled into the car and met Dave, Hilary and Scarlett at the Queen Elizabeth Theatre. A few minutes later Mother and Father arrived from the Victoria ferry.

Our worries were over. The huge auditorium was packed. It was more like an open house than a lecture film, as people I hadn't seen for years came up to introduce themselves at intermission. My last Vancouver class came en masse with their new teacher. As in Victoria, the atmosphere was one of friendliness and the animals helped.

As the Queen Elizabeth held three times more people than the McPherson, we decided that the bear pit, otherwise called the orchestra pit, was the best place for Skookum and Cultus. Mother and Hilary went in with them and answered questions when the children came down the aisle to see the bears play. Father wandered around with Scarlett who was in her element as far as having an audience was concerned but out of character, fortunately, when she conversed so politely. A couple of my former pupils who had hurriedly read everything they could find on coatimundis strolled in the foyer to talk about Pixie and Pete. The coatis enjoyed being cuddled but didn't take much interest in learning about themselves. They slept through the performance. Gypsy, in her night-time fashion, was fast asleep around my neck, looking like a splotch of black ink against my new white mini.

After their busy days the animals were content to relax and leave to the audience the excitement the film had engendered.

The enthusiastic response to *Coast Safari* was an inspiring climax to end a winter of worry.

8
A Fiercesome Threesome

Gypsy is a gibbon
As of course you know
And every time you look at her
She puts on a little show.

Gypsy runs the classroom
As free as free can be
And sometimes if I call her name
She looks around at me.

Gypsy climbs up to the shelf
And knocks off all in sight
And when it causes a very loud noise
She marvels at her might.

Gypsy is at the climbing stage
She scampers across the room
So you'll have to watch just where you step
Or you'll bring her to her doom.

Gypsy is a load of fun
She swings around in joy
But when she is sleeping
She's like an unwound toy.

(Jane Gatt)

Upon her return to Monterey School after the spring holidays, Gypsy was no longer content to stay around her tree on the classroom floor. The whole room became her domain and I wondered if we could last till school broke up for the summer holidays. Or would she have us climbing the walls figuratively as she did literally?

Gypsy expanded her territory by two main avenues: the high-level route by swinging to the top of the tree and running along the window ledge to my desk; and the low-level route by crawling to the nearest table then swinging from leg to leg till she climbed up the back of my chair. The children suppressed their giggles and pretended they were transfixed by their studies but she was hard to ignore. Gypsy would sit right in the middle of my papers, tear out the pages of my daybook (much to the amusement of the class), pick up the most chewable pens and pencils, feel in my drawers for anything that fitted well in her mouth, then throw every paper I was marking onto the floor. The class's giggles turned into audible titters.

But the most horrendous crime of all was when she quietly climbed up the back of my chair and, unbeknownst to me, reached out her long skinny arms and *pulled off my wig*! My face must have gone beet red in embarrassment. Gypsy of course was probably the only one in the room familiar with how I looked at home. Now she had rudely revealed my closely kept, year-long secret. The giggles and titters rose to a roar.

Hair had always attracted her attention, even a single strand that she found on the floor. I just wished she hadn't been attracted to a whole head of it, especially when it was detached. She must have remembered the day when I ran up the stairs to the medical room, fell flat on my face at the top, and looked up to find my wig spinning down the corridor on the janitor's freshly shining floor.

Still, I don't think she embarrassed me as much as the time she tried to slip past my dress and climb inside my bra!

Gypsy obviously wanted to explore the room but not yet on her own. If I left the desk, even to go to the blackboard a couple of feet away, she screamed at her departing security and leaped to whatever part of me she landed on. A four-pound bolt that usually sent me staggering! Being a teacher of the active type and given often to gesticulating, I found her too much in the way when she hung around my neck. If I put her down she flew to my legs with an *oo-oo-oo*, and clung tightly while I hobbled around the classroom, four pounds off balance.

At Island View she expanded her territory in similar fashion. Once she knew how to slide out of her tree house and master the curtain route, she enjoyed brachiating to my bedroom. If Scarlett complained, she just tugged her tail and whisked away. She rode my legs from room to room and then returned to the one she found most fascinating, the library den, where there were screens and projectors, files and papers, books and pictures, tapes and films, to lick and touch and bite and pull.

Eventually she pushed open the front door and found a ready-made gibbon playground–real swings, teeter-totters and picnic tables built for the children who came to the beach. Gypsy became its favourite devotee. On sunny days she always had an audience.

If she saw me heading for the back door she jumped on my legs for a ride to the compound. As soon as the outside animals saw this strange black attachment to one or both of my appendages, there was a bedlam of coos and trills, chitters and kacks, whines and growls, to which Gypsy replied with incessant screams of her own.

She even expanded her use of space in the car–jumping up and down on the floor, leaping over the seats, banging her head on the glove compartment, hitting out at the ignition keys, striking up at the steering wheel and bouncing on my driving arm. A diary entry for such a day reads, "Gypsy and I passed five policemen in the one street on the way to school this morning. I may have to leave her at home."

At this stage in her development she seemed to go berserk in flurries of excess energy, not only in the car but at home and school as well. She scampered over the floor, jumped, swung, pulled, pushed, tore things up, knocked things over, shook her head like an idiot, and all the time emitted soft little hoots.

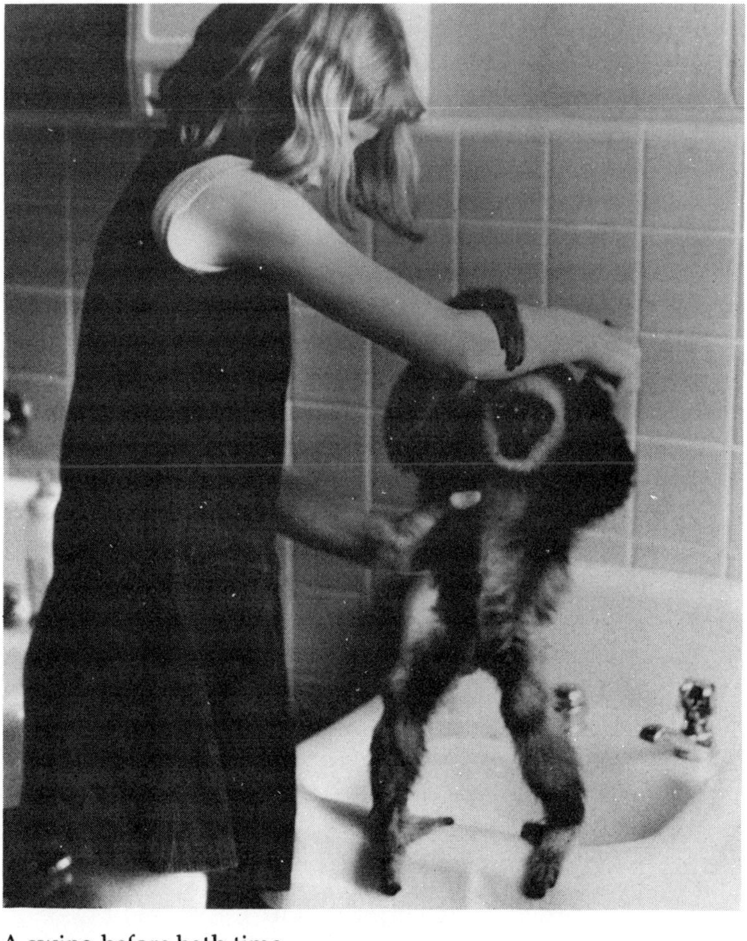

A swing before bath time.

Friends like the Gyes, the Winterburns, the Bennetts and the Sandeman-Allens delighted in taking Gypsy to their basements to watch her swinging on their pipes. Sometimes they tied ropes across their ceilings to give her additional room for acrobatics.

By now Gypsy was walking upright with her arms held high above her head, presumably still for balance. The more she practised, the less she depended on her arms, although they did flail at her sides like two skinny whisks ready to topple anything within reaching distance.

Except for humans, gibbons in the wild spend more of their time upright than any of the apes. Half their waking hours are spent just hanging without necessarily moving from place to place. When they walk upright on branches, they hold their arms high above their heads or out from their sides ready to grasp branches if needed. Wild gibbons rarely come to the ground.

Gypsy, in a world of few trees, spent most of her time on the ground so she learned to walk upright with her hands at her sides. Only when feeling her way on uneven surfaces did she hold her arms above her head for balance.

Some researchers are of the opinion that our human ancestors were walking upright from the moment they left the trees. They cite as evidence that human babies are able to suspend themselves by their arms for as much as a minute shortly after birth. In studying Gypsy we were made more aware of our similarities than our differences.

On the first fire-drill practice of the new school term, Gypsy radically extended her home range. We were all busily engaged in science research when *Rinnnnnnnnnnng-linnnnnnnnnnnnnnnng!* the fire alarm sounded. Instantly we stood up and filed mechanically into the hall and out the door to the playground. From each exit other classes marched in neat, orderly rows to assemble outside the school. The vice-principal and an inspector from the Fire Department were on the steps waiting to see how much time we took. The whole school from kindergarten to grade seven vied with each other to be the quietest and the quickest class to line up in front of the fireman. We stood like soldiers on the parade ground, stiff and still, ready to hear the result.

Suddenly *oo-oo-oo-oo-oo-oo-oo-OO-OO-OOOO-OOOOO!* A frantic screaming reverberated along the walls of the classrooms in line with us.

Gypsy! The whole class remembered at the same time. I broke ranks and ran back into school. The classroom door stood ajar. Now where do you go when you are lost in a school?

Gypsy, abandoned so abruptly, so swiftly, so totally, by everybody she had come to trust, found herself alone in one vast empty deserted room. It was a room she had never explored alone. A room she had seen only on foot–my foot. Fear of the new was swept aside and suddenly confronted by the danger of being deserted, her innate ability to survive came to her rescue. In the unaccustomed silence she sought safety by screaming her way to the door in pursuit of the security provided by her group. On newly mobilized upright legs she took a left turn in the strange bare corridor, a right turn by the medical room, sprawled down the stairs and skedaddled into the office where with many frantic *oo-oo-oo*'s she shimmied up the leg of the astounded principal. He was still in a stupor when I arrived.

Not so Gypsy. At the sound of my voice she made a flying leap to my neck, dug her fingernails into my skin and left tangible evidence of her terror.

In seeking security from strangers in strange places, Gypsy had shown a capacity to withstand shocks and cope flexibly with the challenge of new situations. Still, as Joan said, "Gypsy kept squeaking for all the next period because she was so afraid it might happen again."

Now that Gypsy had been to the office, we had difficulty keeping her in the room. She wanted to explore the rest of the school. A door monitor was appointed to watch that pupils weren't followed when they left the class. The monitor's duties included taking her to fire-drill practice.

I was surprised that Gypsy didn't extend her territory upwards at this time by shinnying up the hot-water pipes to the rods that supported the blinds. From there she could have leapt to the lights. Instead, she became engrossed with the floor. We kept a very clean classroom for two reasons: first, Gypsy was our vacuum cleaner by sucking up nails, staples, thumbtacks, pencils, erasers and chalk (and once after I had mislaid it for a week, my diamond engagement ring); and second, to keep *her* clean, we scrubbed the floor daily to offset the ire of the janitor.

But Gypsy didn't learn to put things back in their places and the sound of objects falling was common in the classroom. One day she knocked over the cans we were collecting for the Fun Fair, then enjoyed rolling them back and forth across the floor. Another day she wound herself up in the Red Cross wool.

She had a fascination for small things. She would sit on the back of a pupil's chair and pick out a hair. After she pulled it out, accompanied by a little squeak that coincided with that of the owner of the hair, she held it between her fingers and bit it.

Sandra Slaght described her interest in insects.

> Gypsy crawls along like a spider with flat feet. She spots something with her brilliant gleaming eyes. She crawls quickly, quietly, to reach it, then slows down as she approaches. Puzzled, her face moves closer, then a second later she gobbles up a tiny black ANT. As rapidly as she can she runs back to touch something familiar before making another foray.

Gypsy liked small animals as well as small objects. The two she found most engrossing were the month-old coatimundis. She didn't care for the more boisterous bears. Sue was now living for the summer at Island View to help with

Well, I just want to have a look.

the chores and run the coffeeshop. She baby-sat Skookum and Cultus while Pixie and Pete went to school. The children prepared new record books and began to note down their observations on the coatimundis just as they did with Gypsy.

Wendy Fischer described her first meeting with the coatimundis:

I came into the classroom
Feeling full of awe
I was quite surprised
With what I saw.

Two coatimundis
Creatures from the south
With ramrod-straight tails
And long, pointed snouts.

My first impressions were
"My! How small they are!"
But when I saw their tails, I thought,
"How long they are!"

They made a soft little churr
With a very high tone
But those two little beasts
Sure weren't alone.

Yes, they had Gypsy
Who ran around
In curiosity
And knocked them down.

So it was Gypsy, Pixie and Pete
What a threesome they are!
I just wonder how they act
In Mrs. Hancock's car!

The same as they did in the classroom, Wendy! But I think you say it better:

> Round and round the classroom run Gypsy, Pixie and Pete. Under the desks they climb on your feet. They squeak and squawk and utter strange cries. But, oh, they're so sweet. Pixie tries to grasp the feeding bottle but it is too round and shiny. Pete lollops round the room crawling up people's legs but when picked up, cries. Gypsy tries to take the bottle from both of them.
> Then when feeding time is over the three animals are ready for another game. Chase? Yes. And they're away. Gypsy tries in vain to catch one of the coatis. But Pixie and Pete are too smart. They crawl under the teacher's desk and get up on the seats behind the pupils. They play with our pencils and try to bite the lead. They worry the straps of our school bags and play with our shoelaces. Pixie and Pete brush past the bookcase and all the magazines fall out. Gypsy knocks down the pole used for pulling the windows open while the coatis crawl and delve into anything that happens to be on the floor. They run around and enjoy themselves for, after all, they're only young once.

Marion Condrashoff's version was in verse:

There are two Bandits loose under desk seat three
Gypsy will catch them, wait and see,
They're under arrest for disturbing the peace,
They don't know where to go, what a couple of geese!
They have little masks fastened tight on their faces
They'd better get going and find some hiding places
For Gypsy is scuttling along the floor
And when she catches them, she's going to get sore!
So come on now, get going–beat a hasty retreat
Run as fast as you can on your small padded feet,
Hop on a case or jump on a seat
You're not Bonnie and Clyde, you're just Pixie and Pete.

Believing that for children, as well as adults, education can come through entertainment, I didn't worry if some of my colleagues looked askance at the chaos they described as coming from the grade six classroom. Learning still ensued. I was glad to know that the students agreed with me. Wendy Fischer was one:

> While we sit in the classroom working, there are a lot of things happening. When the two coatis play together it is very funny. They always try to bite each other's tail but seldom do. Usually they bite their own (ha!) then speed away. And when I say speed away, I mean just that. They scamper across the room like lightning, and boy, can they go! Then of course Gypsy gets her eyes on them. She sneaks up on them very stealthily then takes a swipe at them. After that she runs out of their reach. You might say she plays her hit-and-run game. She really wants attention, you know. Then the three animals play a hilarious game of tag. One of the coatis whirls around the leg of a table, followed by Gypsy trying to grab its tail. On and on they play till they're exhausted. Late in the morning they finally sleep altogether in the box on the classroom floor. It's fun to be in a classroom with so much going on and

yet I can still learn a lot from those three fascinating creatures.

Fortunately, the principal must have thought so too as, during those last three months of the school year, he tried to persuade me to stay on the staff. As I needed a regular job with a guaranteed pay cheque to support operations at Island View, I didn't want to resign. But David decreed that my time would best be served at home. He warned that when school resumed in the fall I would be in Alaska.

Nurturing our orphans fulfilled certain needs in us all. One day the principal showed me part of a psychologist's report on one of the students who had been in my homeroom class the year before and was now in my English class: "Marilyn Dowler is tall, dark, slim and wears glasses. She wants to be a vet. *She enjoys working with the school ape.*"

Surely a unique sentence in any official document! Although Marilyn was not the academic type and had many social and emotional problems during those two years, she was serious, industrious and one of my most reliable helpers. She may have had to strive over spelling and math but when stimulated by the tortoises, barn owls, stranded robins, a gibbon and the coatimundis, even academic subjects came more easily. Now that Gypsy's social space had so rapidly expanded, Marilyn took turns with Dorcas as my chief assistant.

With the coming of the good weather the class no longer stayed inside at their desks to eat their lunches. At noon they dined on the lawns. The children sat around Gypsy, a goddess surrounded by her servants, and served her tidbits from their supplies. Gypsy still liked her bottle but she also ate oranges, apples, celery, bananas (at last), carrots, boiled eggs and the ubiquitous baby cookies. She also ate staples like peanut butter sandwiches when lunching with the class and cheese when lunching with me.

Marilyn kept a watchful eye on Gypsy's diet and growth.

Sometimes Marilyn's mother brought her out to Island View for the weekend to work with the rest of the animals and expand her experiences beyond the "school ape."

With every great leap forward in her development, Gypsy demonstrated an unexpected return to a period of dependence. Perhaps she needed extra comforting at a time when she was decreasing her independence in other ways. Perhaps she was jealous of the attention I had to give to the other animals.

She clung to my legs as I hobbled around the rooms or shovelled garbage from the compound, hung on my neck while I tried to cook, climbed on my back while I leaned over the tub, or curled on my lap as I sat at the typewriter. She was content to play on the keys with her feet till about nine o'clock at night. Then she sneaked off my knees and surreptitiously crept into the bedroom, jumped on the bed, pulled down the covers and climbed between the sheets. I followed at midnight. It took a harder heart than mine to remove her from the bed and put her in the living room pen. I usually crawled in beside her.

For the first few nights she stayed snuggled under the blankets. Then about four o'clock in the morning she began an annoying habit of waking me up with a kiss. If I didn't respond she tugged at my hair, then shimmied up the curtains to sit on top of the closet. If she stayed there I could have gone to sleep. Instead she pelted me with sweaters. Despite a succession of violent temper tantrums at not getting her own way, my heart hardened and I put her in her own bed in the living room.

As winter activities came to an end and summer activities were about to begin, Scarlett, once the lord of the living room, also began to throw temper tantrums for attention. One morning while Dave was at home she waddled in to the bed, hooked her way up the covers and sneaked quietly down between the sheets to the bed posts. I awoke to her muffled growls as she tried to make up her mind which set of toes was mine. "Dave," I whispered in terror. "Scarlett's at the end of the bed. Get her out before she draws blood." I was still smarting from the night before when she brazenly lunged at me from her smug perch on David's shoulder and nearly poked my eyes out. Gingerly running his hands down to the end of the bed, feeling for her feathers, and talking in mollifying tones, "Here, now, Scarlett, good bird, Scarlett, come here, Scarlett," David manoeuvred her to his side by the wall. Meanwhile I surreptitiously withdrew myself and moved out.

Another time when the door was closed but the room empty, she vented her frustrations in other ways. When confronted by a door, invariably Scarlett was polite. She always tapped several times with her beak on the woodwork and said "Hello" or "Hi, there" or "Yoo hoo." But if she heard no answering invitation she revealed an alarming lack of manners, pushed the door open with her bill and sneaked inside to wreak havoc. This time she shredded the tongue and laces of my hiking boots, tore holes in my new leather shoes, ripped off the handles of my handbag and made a green carpet yet greener with copious defecation.

As her home on the TV set was situated in the corner beside the bedroom door and Scarlett always waited for me with a hiss and an outstretched open beak, to sneak past her I developed a routine sideways wiggle of my bottom and a wag of a towel. Even when she wasn't at home I retained the habit of that oblique scrape-by-between-rooms mannerism.

The TV set, at different times home and hiding place to otters, bears, raccoons, coatimundis, a hyrax and a gibbon, had long been defunct for its original purpose. Though it remained a den for Gypsy, Skookum, Cultus, Pixie and Pete, I devised a new perch for the pesky macaw. With Hilary's help, one weekend I hung a branch from a hook in the ceiling and hoped that Scarlett would remain there in her space rather than intrude on everybody else's.

But Scarlett was not to be fooled. She found she could

sway back and forth until she had gained enough momentum to swing across to either the curtains, the chesterfield or me. Once having gained a beaking or footing, she was down on the floor in a flash of her varicoloured feathers.

My only successful innovation was to replace the metal edges of her food tray with rubber tubing to deaden the effect of one of her most annoying attempts for attention. She liked biting at the metal with her beak, which to me was like listening to a fingernail or piece of chalk scraping and squeaking its way over a blackboard.

Fortunately, summer sent Scarlett out to her totem-pole perch by the front door where on weekends she was now joined by Gypsy and Pixie and Pete.

At this stage of exploration the trio never strayed further than the swings and picnic tables, and under Susan's surveillance didn't get into trouble any greater than grabbing a few sandwiches. All being animals of tropical origins from places far away, they did look a bit incongruous on a British Columbian beach. The bears looked more at home but as they were likely to grab fingers as well as sandwiches, they only went outside on a leash. Susan would take them for a stroll when most of the people had gone.

Scarlett had long monopolized the attention of visitors to Island View Beach. Now she had to share her audiences as well as their peanuts and ice cream with one gibbon, two bears and two coatimundis. Perhaps that was why she strove for greater heights.

One Sunday she climbed from the totem pole to the television aerial then continued along the roof to perch on the chimney. Bored by the lack of entertainment so high above her audience, she tried to come down, thankfully not by way of the chimney. Either her good memory failed or else she sought a more challenging method, but she made no attempt to descend by the aerial. Instead she tried backing over the edge of the roof, as she was accustomed to do over the side of the TV set. But the ground was many feet further down than the living room carpet. Next she tried coming down head first. But keen eyesight prevented her from fluttering beak first into hard cement. Well then, why not walk a tight rope across the electrical wires, over the transformer box, into the oak tree and use its branches to get down to the lawn? Besides, she could have some fun on the way.

Scarlett strutted along the wire and entertained a quickly gathering group of amused spectators. She looped and leaped and pirouetted, an amazing display of agile avian tight-rope gymnastics that rivalled Gypsy's acrobatics. She not only danced but she also sang her complete repertoire of throaty vocalizations. I came along while she was still halfway across the wire and before she had gone too far in her choice of words.

To the disappointment of the crowd, I climbed a ladder and offered her the end of a broomstick. This was the only way I could carry her around without too much damage to myself. I'd say, "Come on, Scarlett, there's a good girl." Echoing, "Okay, come on, there's a good girl," Scarlett would step onto the stick and immediately strut to the end I was holding. And so we'd move, with me rapidly changing ends like a juggler to keep out of the way of her constantly lunging beak.

This time, to my surprise, she grabbed the broomstick with her beak, swung her feet forward to latch on more firmly, then whispered ingratiatingly, "Come on, come on." I descended the ladder in haste, while she waddled along the broom towards me. She reached me just as I reached the bottom rung, then she gave me a sharp peck for my interference.

Gypsy now paid more attention to her new companions, Pixie and Pete, than to Skookum, Cultus or Scarlett. But she must have been listening to the macaw because to my astonishment she started imitating many of Scarlett's vocal overtures. Her hoots and whoops now bore the unmistakable tones of screeches and squawks. Beach-goers, already confused by animals swinging in a children's playground, by southern species in northern habitats, by jungle denizens that weren't in cages, must have been further confused by such crossing of the language barriers.

I had no time to throw light on their confusion. I was busy in the compound. The swallows were choosing nest sites on every window ledge. The pigeons were laying eggs in the food preparation room. Wynken and Blynken the barn owls were sitting on another clutch of eggs, something they had been doing continuously since January. A female mallard that had come in mangled a month before was now patched up and happily mated to a once-forlorn male that had been trying to break up a pair now sitting on a clutch of eggs by the pump house. While they were all trying to produce families, Pierretta the Canada goose had just taken over the mothering of six orphan mallard ducklings.

A couple of puffins had burrowed into a hole in the sea-bird pen and promised to be the first of that species to breed in captivity. Baldy, the one-winged eagle, had set up house-keeping with another one-winged eagle, and though they had not yet produced an eaglet of their own, they were raising an orphan. And the peregrine falcons were once again looking as if they should lay. While we were waiting for them to provide their own, an official from the Fish and Wildlife Branch brought from the Queen Charlotte Islands a nestling falcon that had been confiscated from poachers.

Sometimes while I was working in the compound Gypsy heard my voice from her play on the front lawn and whooped around the side of the house to find me. I never failed to answer her appeals as she stood by the fence asking to come in. If the bears joined her I had a merry and complicated time trying to unravel enough legs to do the chores, as they all tried to climb up my jeans. Sam the seal and Pierretta the goose always came over to cough and honk and see who had invaded their territory.

On rainy days and nights I had them all inside the house.

The coatis, still tiny enough to drink only half an ounce of milk at a feeding and dependent enough to climb my shoulders for their afternoon siesta, developed socially faster than even the bears. They poked their long, inquisitive snouts into everything: chewed busily on the woodwork, found their way into Gypsy's tree, climbed the drapes, balanced on the curtain rods, tried out the stove, and cut through the phone cord. I cringed at what they would be like when they reached, as the books said, "the size of cocker spaniels."

I was obviously partisan. My diary at the time reads: "Drat that macaw! She makes such a mess, tipping over her food tray and knocking two cups to the floor." . . . "Bears can get themselves into so much trouble, tear everywhere, climb the frig and toss everything from the top." . . . But "Gypsy looks so beautiful brachiating along the curtain rods, dangling by one arm and doing pirouettes better than any ballet dancer."

My nerves were less fractured when we went visiting. Or else my diary echoed my party manners: "After a day with Kerry and Lyn Joy at their house in the Highlands, Skookum cut his pads on some wire so she sat on their chaise longue on the lawn with all four feet up in the air and crying pitifully. What an expression! Her tongue lolled sideways out of her long snout and she looked so stupid yet so lovable."

The bears were in better spirits one weekend later when we took them with Robin and Jill Bennett to Little Darcy Island, my favourite Gulf Island and a photographer's paradise. It rained but that didn't dampen Cultus' spirits. No sooner had we pulled our rubber boats up the beach than Cultus charged up the first tree right to the top. He came down more slowly, trying out different methods for different branches. The one he miscalculated almost brought him to the ground–flying. He teetered to the end of a slowly bending side limb, turned and tried to get back.

"He's going to fall," yelled soft-hearted Robin.

"He's hanging by one claw," observed David.

It was true. Cultus was dangling from one paw like Gypsy did but not with quite the same poise.

"He's drooping... he's dropping," sympathized practical Jill who ran under the tree ready for the rescue.

But somehow quite miraculously Cultus averted his downfall by gripping his claws tightly into the bark and swinging his fat little body back to the limb. The branch bent under his weight but didn't break. I stood open-mouthed, looking up, irate because I had forgotten the cameras.

Cultus then chose a branch more his size and bounced and jumped along it, seemingly unconcerned by his entrance into parachuting school. Attracted by his high spirits, Skookum followed him up the tree.

"Watch her," I worried. "She's the scared one. She'll get up partway, think that it's easy, then won't be able to come down."

Perhaps she suffered from feelings of inferiority in the wake of her more boisterous brother but as I prophesied, poor Skookum got as far as the branch Cultus was using as a trampoline, tried unsuccessfully to bypass him, then clung to the trunk of the tree, moaning like a human child. Jill helped her down. Cultus didn't descend till he saw us on the beach launching the rubber boats for our departure.

In the last week of the school year, activity is supposed to wind down. In my case life continued on its usual spiral of never-ending climaxes. Gypsy learned to climb the hot-water pipes and added the ceiling to her final exploration of the classroom. The coatis clambered in and out of the cartons being packed for the time I might start teaching again. All of them played tag with a reel of film being threaded vainly into the movie projector by one of the children. At a time when the door was often being opened and closed by people on various end-of-the-year errands, Gypsy sometimes flew out to visit lunch-time companions in other classrooms, while the rest of the merry trio scampered in pursuit.

I made a mental note that our little ape would have to be a dropout in her first year of school, not because she was too backward but because she was too far advanced.

At home the phone kept ringing with requests to pick up a beached seal or an oiled seabird, to mend the wing of a hawk or disentangle a gull from a mesh of wires, to lend books, give talks or eradicate four-footed intruders from houses where people believed that one woman's pest should be another woman's pet.

Not only animals but people kept coming. Animal technicians wanting to talk. Artists wanting to paint. Doctors wanting blood. Students wanting work.

Mr. and Mrs. McKeever, avian enthusiasts from Ontario, just wanted to look at owls. Blynken gave them a back-slapping welcome and a clip on the face as they entered the barn-owl enclosure; and Gypsy, at a time when she was between diapers, splattered freshly gorged cherries and milk down Mr. McKeever's neck. While I moaned in anguish, my mother-in-law cleaned him up and Mrs. McKeever murmured comfortingly, "There, there, it's only a small patch, dear."

Meanwhile, amid the chaos, preparations for the summer field trips continued: to the Channel Islands in California to study seals and sea otters, to the Rockies to study ptarmigan, to a Pacific coast island to study seabirds, to the Yukon and Alaska to study mountain sheep.

There was no letdown in excitement even on the last day of school. I wondered for a while if I would get there. Susan was out riding her horse. David was away in Barkley Sound on the west coast of Vancouver Island looking at eagles. The bears, thankfully, were locked in my bedroom for the morning. I didn't intend to take them to school even though the principal had at last accepted my letter of resignation. I was in my dressing gown feeding the coatimundis when *crash!* the trilamp came clattering to my feet. Gypsy, who was using it for jumping practice, rode it down all the way. It might

Good morning, sir, this is the Siamese Strangler speaking.

have stayed intact instead of breaking into a dozen pieces if the vacuum cleaner had been out of the way. Gypsy scooted up my leg but the coatis vanished at the first sight and sound of the smash.

I spent the next hour looking for them. Realizing I would be late for school on the last day, I phoned Irv to give the reason. He groaned and said something like "What's new!" The coatis were still missing by the time I imagined the bell rang so I decided to dress, lock all doors and go without them. I could clean up the mess when the year was over. I reached towards the closely packed hangers in the living room closet and *ouch!* my fingers made contact with two rows of sharp pointed teeth, not what you expect in the average wardrobe. As I reeled back in surprised pain, two pairs of bright little eyes were shining from the shirts.

The coatis in the clothes cupboard was an apt anecdote for my speech at the school's farewell luncheon. In his address, Irv Dawson recounted wryly some of the highlights in having an animal nut for a teacher . . . tortoises underfoot and apes in the air, parrots answering the phone and bears bawling in the background, children lunching with owls, parents making diapers instead of concert costumes, and teachers with the oddest excuses for getting to work late.

"Having a Hancock on the staff is not without its hazards," Irv grinned ruefully, remembering fire-drill day. "She doesn't send notes to the office, she sends a Siamese Strangler."

The children gave me another party that afternoon. Amid the confusion of cartons and coatis and a gibbon who kept eating my flowers, they presented me with a special farewell gift of their own.

Sitting in a wicker basket on a green cushion, wrapped in blue bows and very much alive, was a SNOW WHITE RABBIT!

9
Prestigious Parrots in the Yukon

When I arrived home from school that afternoon the bedroom was empty. The bears were gone.

"I left them in the compound," said Susan, a trifle subdued as she responded to my shouts by hurrying in from the store.

I looked out the window of the living room to find the big, heavy gate swinging open. "They must have got out," I panicked. "Susan, you go down to the potato fields and I'll go down the beach."

Leaving the store to be minded by one of the boys in the cabins, I ran down the boat ramp. "You haven't seen a couple of bear cubs walking along the beach, have you?" I asked desperately of the first group of picnickers.

"What?" they repeated in unison. "Alive?" Then they laughed and said, "Oh, you mean a couple of *beach boys*, eh?"

"No," I groaned and kept on walking.

As nobody in the next half an hour had seen either lost bears or lost boys or lost boy cubs, I knew that Skookum and Cultus hadn't gone to the beach. Neither had they gone pattering through the potatoes. But they had tried playing in the toolshed and that's where we found them, curled up in each other's short, stocky arms on a pile of sacks. They had never left the compound. I had just jumped to the wrong conclusion.

The next shock for the day was a visit from the Sandeman-Allens to say that their daughter wasn't allowed to stay for the summer to look after the coffee shop. It seemed that Susan had bought a horse against her family's wishes.

In an animal-conscious community like Victoria there was no lack of volunteers to help with the animals. Mother would be staying at Island View for the summer but she needed assistance and I could think of no one better than Susan.

A teenager called Allan was the next on my list of volunteers. However, when he let the tortoises out on the lawn and then lost them while he was looking at them, I wondered what the animals might do if he wasn't looking at them. I decided Allan had to go.

Luckily for us all, he was replaced by David Litt, a cheery, responsible student who was practical and hardworking enough to survive a summer at Island View. With his assistance, the animals thrived and Mother catered to the customers at the store. David even lent a hand when the water ran out, not from too much precipitation, as in the winter, but this time from too little.

With the unusually hot, dry weather, the springs dammed back by the farmers for their potato fields and a leak in the line bringing water into the house, it was back to a pattern of bathing at the neighbours and bathrooming in the public park facilities.

Our field assistants that summer were Errol Anderson, a young naturalist from Vancouver, Grant Blundell, a raw recruit on his first summer job but one of the most easy-going and efficient teenagers you could meet, and Bob Flather, an older man from Santa Barbara, a competent outdoorsman with degrees in commerce and engineering as well as zoology. For the four of us the summer brought water in excess, as first we were drenched by waves in a rubber boat off the Pacific coast and then we were flooded by rain in a tent on the northern tundra.

One problem that I was glad to find solved was Cultus and Skookum. The bears would grow too big for our resources at Island View. For their own protection they couldn't be released in the wild. Zoos in Canada, accustomed to receiving spring orphans, were not accommodating, so a good home for Skookum and Cultus was found on a game farm in Kentucky.

Gypsy was my special worry. She must have sensed my concern because for the weekend prior to our departure she refused to let me alone for a minute. She followed me everywhere, to the bathroom to wash the laundry, to the bedroom to pack the clothes, to the kitchen to prepare the camping gear, to the living room to do some last-minute sewing. She stayed so close she got caught up in my thread and I almost packed her with the luggage. She screamed piercingly if I left her. I knew then that if ever I had an infant Homo sapiens I could never leave it crying in the crib as I had previously theorized. My child would probably be one of the most spoiled in the world.

As Mother and David Litt would never be able to get any work done with an ape clinging to them all summer, Bob and Errol helped me prepare a spacious enclosure in the compound for Gypsy. A similar one was prepared beside it for Pixie and Pete.

There were bars to balance on, ropes to swing from, branches to play with, logs to chew and plenty to eat and drink. Both pens were built high to afford good vantage points over the other animals.

I sat with Gypsy in her holiday home for hours but she wouldn't even look at the new playthings. Instead she clung tightly to me with her thumb in her mouth. We all took turns sitting in with her but she showed no curiosity to explore. She must have sensed that for the third time in her life she was going to lose her mother.

She had a grandmother because on several occasions I had left her with my mother-in-law, but though on familiar terms with her, Gypsy was not particularly friendly to her at the time.

Mother now climbed into the pen. While I sneaked away to prepare some lunch she cuddled the little ape and tried to distract her attention. Then we all lunched together inside the enclosure with the coatimundis. Gypsy wasn't even interested in Pixie and Pete. She just clung to my neck and sucked her thumb.

David was getting impatient. We had to go. The boys had finished loading the truck. Feeling like a bad mother abandoning her baby, I slunk away.

Two weeks later we arrived back from California in the middle of the night. Not wanting to wake everybody up, we slept in the back of the truck till morning. Father came out at breakfast time to tell us that there was no water. And Mother came out with Gypsy. I clambered over the sleeping bags expecting her to leap into my arms with an excited squeak. Instead she turned her head away and ignored me completely. While I was worrying about her feeling rejected, there she was clinging to Mother's neck and freezing me with a disinterested stare. Glad that she had been flexible enough to adapt to a new situation, I left her to her new mother and prepared for another trip the next day.

This time it was to Triangle Island off the northwest tip of Vancouver Island to study seabirds and sea lions.

Back again at Island View for a day before the next expedition to film grouse up Island, I was happy to see that Gypsy, now attached firmly to both Mother and David Litt, had forgiven me for deserting her. She still didn't leap into my arms but she no longer ignored me.

She had plenty of companions: the tortoises which had been discovered under the house, the domestic kitten, the white rabbit, and Pixie and Pete who were almost as acrobatic as she was. Gypsy mothered them all. During the day they played together in the outside enclosure. At night Gypsy was treated preferentially and came inside the house to sleep with Mother.

As soon as a sick albatross was doctored and released, the rest of us were all away again, this time to the north on a more extended trip.

Errol wasn't coming this time. "Let's take Scarlett," David suggested impulsively as Grant and Bob and I crammed into the crowded cab of the truck.

"You've got to be joking," I retorted in horror. Even the boys looked askance. "Travel in the truck with Scarlett? That neurotic macaw attacks me any chance she gets and when you're around, everyone else as well."

"If we make it to Carcross she can meet Polly, the famous Yukon parrot. We can film them while they swap stories of life in the last hundred years."

And so Scarlett went north. She didn't meet Polly but she had plenty of other adventures. And in seeing how she scraped through some of those escapades, I realized why parrots live so long.

She had to give up her favourite foods like T-bone steaks, spaghetti, chicken and whipping cream and be content with mere bird food like sunflower seeds and walnuts. She had to stay inside the truck with the engine running continuously when days were cold and it snowed. She had to put up with mosquitoes whose reputation is legendary in the boreal forest. Worse still, she had to put up with me.

It was nerve-wracking as well as a tight squeeze with us all in the cab of the truck. David drove and I sat in the middle with my legs straddling the gear shift. Grant or Bob, taking turns in the front, sat by the door. Scarlett perched on the narrow back of the seat between the radio which blocked her loving intentions towards David, and the tape recorder which blocked her unloving intentions towards me. I sat bolt upright. If once I relaxed and slumped against the back of the seat, Scarlett seized her opportunity to step over the tape recorder onto my shoulder.

I would freeze, close my eyes and face straight ahead. The pernicious macaw would then rub her head against my cheek, sway from side to side, ruffle her feathers, croon softly in imitation of some lullaby, then whisper seductively, "Hallo, I love you."

"Look at her, Lyn, she's just trying to be friendly," laughed Grant from his safety zone on the other side of the tape recorder.

"Okay, Grant, then you be the one to take her off me," I suggested icily through gritted teeth, not daring to look anywhere but at the road ahead.

"Not on your life," he grinned. "When Dave's around she hates me almost as much as she does you."

So many times I was beguiled into believing Scarlett wanted to make friends and call a truce! Vain hope! Without warning, right in the middle of a lullaby, *wham!* she raked down her powerful beak and blood flowed from the slice in my chin. She scolded, "Bad bird, bad bird!" then followed a string of incomprehensible imitations of human speech as she thumped her beak against the back window to relieve what supposedly were frustrated jealousies.

To give her the attention she craved, David would occasionally allow her to perch on his shoulder while he was driving. But not knowing this was a special privilege, Scarlett would reach out with her beak to catch the steering column, rather a hazard when trying to turn corners. Despite her loud shrieks of protest, she was soon returned to the space between the radio and the recorder.

She spent her nights perched on the steering wheel with an ample supply of newspaper spread underneath. Invariably, next morning we found she had torn up the paper and spattered the gear shift green till P for Park, N for Neutral, D for Drive and R for Reverse were totally unrecognizable by the end of the trip. And long after the expedition, we were still finding nuts and sunflower seeds in the most remote recesses of the truck.

To keep balanced, Scarlett would have to spread her wings and grip the back of the seat more tightly in rhythm with the bumps. While thus occupied she was distracted from paying attention to me.

If ever we left the truck to look at some animal or film a scenic view close at hand, she would waddle over to the window, climb out onto the rearview mirror and call, "Good-bye, good-bye." Then she'd flap her wings, bob her head up and down and yell "Yoohoo" to passing traffic. There were plenty of double takes but luckily no accidents.

If David walked away out of sight she'd hoist herself higher by trying to claw a foothold in the smooth, shiny surface of the roof. When she ran out of roof she swung up and down on the bamboo grouse noosing poles which protruded from the end of the truck, or if she wanted an even higher vantage point, she'd climb onto the rubber boat and the outboard motor above that.

One day we dashed back to the truck and forgot she was still on the roof. We had driven a couple of miles down the road when I suddenly yelled, "Where's Scarlett?" The brakes burned to a halt and we jumped out to see Scarlett looking like the Red Baron, clinging with both feet to the propeller of the outboard, flapping her wide red wings in the wind and screaming, "I'm a parrot, I'm a parrot." Surprisingly none the worse for her wild ride in the wind, she beak-footed her way down to David's shoulder and breathed, "Hallo and how are you?" in husky, loving tones.

She survived mishap better than we did. Another time we were bumping along the Alcan south of Whitehorse when the back end of our truck suddenly shuddered and lurched sideways. And there, looming broadside in the driver's window, was our trailer. The front left tire had blown and slowed the truck but the trailer continued to travel at the same rate of speed and swung past us.

To apply the brakes would have caused a catastrophe so we screamed across the road to the left side. Miraculously, the truck stayed on the road. I froze in my seat as the trailer recovered from the slipping gravel verge just in time to avoid our descent backwards into the ditch. Then it swung again the other way, pulling the truck with it. Two of the back wheels of the four-wheel trailer skimmed the edge on the other side. With just enough momentum and weight, the truck pulled the trailer forward again. Seven times the trailer swerved to each side of the road as David strained to steer around every bend. Surely, I thought, those spare barrels of gas will ignite if the long trailer succeeds in pulling the truck into the ditch. I was terrified by the thought of burning beneath them.

But somehow David averted disaster and brought us to a stop. For a full minute we lay back against the seat, our eyes closed, our muscles quivering. Even Scarlett was silent.

Some friendly Yukoners who had braked their own vehicles up ahead and started praying as soon as they noticed our fantastic gyrations, swarmed towards us. One commented as he passed a bottle of Yukon Hootch through the window, "Here, have a drink. You sure need it. That was the closest thing I've seen in a long time. You're all lucky to be alive." Under the influence of a short slug herself, Scarlett regained her usual composure and started singing.

When Grant or Bob was driving, Scarlett sang often. She would sit on Dave's shoulder and dehusk peas while he rubbed her white cheek patches. By the time we reached the Yukon I had gained a little confidence to lean back against the seat but I certainly didn't trust her overtures. Not even when she sat on my shoulder, preened her scruffy feathers casually and whispered "sweet nothings" in my ear. As she laughed gently and made her kissing noises, Bob would say again, "Look, she's only trying to be friendly."

But it was always a prelude to being suddenly pecked on

the cheek. And she darn well knew what she was doing. Dave remonstrated with "You bad bird. Stop that," and he'd cuff her with his hand. But she'd just imitate his words, knock the radio over and for five minutes sulk with unintelligible gabble which must have been a string of curses against me. Then she'd bide her time for another unexpected onslaught.

As we followed the dust, the mud and mosquitoes northwards, Scarlett kept cleaner than we did. She stayed in the truck while we smeared ourselves generously with repellent and toted soap and towels to the river. We wondered why we bothered when our feet squelched into the typical muddy glacial silt of many northern rivers and we emerged even blacker than before.

One day Scarlett met another adventurer, Stephen Jones from New Jersey, who was pedalling a bicycle from Fairbanks, Alaska, to South America. His bike and saddlebags as

If you're going to South America, I'd recommend flying, was Scarlett's advice to the cyclist from Alaska.

well as his person were covered with black muddy slime from some experimental oil the highway crews were using on the road surface. He was as intrigued to meet a South American macaw on the Alaska Highway as we were to meet a cyclist on such a long journey. As we snapped a picture of Scarlett sitting on his shoulder he said wryly, "Good thing I cleaned up at the last gas station, isn't it?"

Scarlett added to her list of admirers when we stopped at Klukshu. The Indians from Champagne return here each summer to their ancestral hunting grounds to smoke fish for the winter. We spent a day at Klukshu watching the Indians trap thousands of salmon by the traditional method of directing the fish into box traps which blocked the spawning stream. While we wandered among the sod-roofed log cabins and caches on stilts and filmed the strips of salmon drying on racks by the river or splayed to cure in the smokehouse, the Indian children followed to stare at Scarlett. By the roundness of their eyes and their rapt expressions when the macaw climbed up their pant legs and sweaters to perch on their shoulders, I was sure none had seen such a multicoloured bird before. Their eyes were even rounder and their mouths more agape when they learned that Scarlett was more than fifty years old.

Since our arrival in the Yukon no day had passed without somebody stopping us to ask if our Scarlett was the famed Polly of Carcross. But these Indian children, scarcely a hundred and fifty miles away, had never heard of this legendary macaw. So I told them Polly's story.

Polly of Carcross was brought to the Yukon back in the early 1900s by her owner, Captain Alexander, an Australian who sought his fortune in the Klondike Gold Rush. During the winters when the captain left the mining camps near Atlin to whoop it up in Vancouver, Polly was left at the historic Caribou Hotel in Carcross. In 1919 the C.P.R. vessel, *Princess Sophia*, foundered on the rocks of the Lynn Canal near Skagway, Alaska, and sank, making it the worst marine disaster in the history of the northwest coast. The entire 343 persons aboard, including Captain Alexander, were drowned. From then on Polly was a permanent resident at the Caribou Hotel.

She became world famous and was known to sing opera, drink rum, speak in Italian and swear in several languages. People said that when someone asked her what everyone asked a parrot, "Polly, want a cracker?" back would come her stock reply, "Go to hell!" One of the expressions she picked up from Captain Alexander was "bread and jam, my dear," which in typical parrot fashion was repeated ad nauseam till it was reported the poor captain nearly went crazy. According to rumour, this famous Yukon macaw was "the oldest, meanest, dirtiest bird north of the 60th parallel."

At this point in my story, Scarlett, still happily snapping sunflower seeds in her admirers' ears, broke in with her variation of Prime Minister Trudeau's "fuddle duddle." As she had once embarrassed us at a gas station stop by addressing a long string of such obscenities to the pump attendant, I hastily engineered silence by stuffing her cheeks with salmon. The children grinned.

Seeing that they were totally fascinated by Scarlett, I plunged into another parrot story, about the famous Louis of Victoria who like Polly of Carcross was also well over a hundred years old.

Way back in 1884 a shy spinster named Victoria Jane Wilson bought Louis to add to her aviary of fifty pet birds. Louis lived a life of luxury with his mistress in a rococo mansion noted for its carved staircases and elaborately moulded Renaissance-style ceilings. In 1949 Miss Wilson died and Louis inherited her fortune.

For the next seventeen years the pampered parrot continued to live in his mansion in the care of Miss Wilson's faithful Chinese gardener, Wah Wong. After Wong's death in 1966 Louis enjoyed the Parrot House, as the mansion was commonly called, for a further five years. Then, despite attempts by several nostalgic Victorians to save it as a historical and architectural monument, it was torn down and a hotel and restaurant built on the site. The rest of the estate was distributed to charities although Miss Wilson's will had specified that this was not to happen till Louis died. I assured my audience that according to the latest reports from Miss Wilson's lawyers, Louis was still alive and well in Victoria.

"You should take Scarlett to meet Louis when you get back home," one of the children suggested.

"Wish I could," I laughed in reply. "But Miss Wilson's lawyers are scared Louis may be kidnapped so they're keeping him secret."

"Wouldn't it be neat to put Scarlett and Polly and Louis together?" another of the children said with a smile. "They'd sure have a lot to talk about!"

Scarlett, still strutting from shoulder to shoulder, bobbed her head up and down as if in agreement with the plan.

"Well, we can take her to Carcross to meet Polly anyway," I replied. "And I'll tell you all about it when we get back."

It was time to go. Dave packed the cameras and folded the tripod. Scarlett pecked at her last piece of salmon and nibbled her last ears. Responding to our mass movement back to the truck, Scarlett squawked raucous good-byes to the village and we continued northwards.

Unfortunately I could not keep my promise to the macaw's admirers as Dave changed our itinerary that summer to exclude Carcross. When I did get the opportunity to visit the Caribou Hotel it was three years later and Polly, at the reputed age of 126, had just passed away.

As befitted such an old Yukon pioneer, she was buried with full ceremony in the picturesque, white-picketed Carcross cemetery. At the funeral another famous Yukon pioneer, Johnnie Johns, sang the macaw's favourite song, "I Love You Truly." The plaque on her grave, surrounded by a small mound of earth, green fernery, plastic flowers and some brown feathers, read:

POLLY

Born 1850 Died 1972

Under this sod lies a sourdough parrot
Its heart was gold, pure fourteen carat.
Polly now can spread her wings
Leaving behind all earthly things
She ranks in fame as our dear departed
A just reward for being good-hearted.

At that time the Caribou Hotel had two other parrots, Polly II, a New Guinea lorikeet sent by a radio station in Montreal, and Polly III, a 45-year-old red-and-blue macaw just like ours. Coincidentally, Polly III came from Victoria and her real name was Scarlett. Furthermore, she had once belonged to a sea captain until World War II when the ship sank and her owner drowned.

It was uncanny how closely the two Scarletts resembled each other. As I walked into the hotel in Carcross, I could have sworn I was back at Island View. A horse poked his head in the window. A cat jumped up on the table. And Polly III, alias Scarlett, welcomed me with a throaty "Hallo."

A notice on the wall made me feel even more at home.

> Unlike most people, Scarlett prefers to have nobody listening. She talks to the TV in the evening and usually enjoys providing an extraordinary range of jungle sounds as a background to the wildlife programs. She barks like a dog, meows like a cat and swears like a sailor . . .

The only difference between the two Scarletts seemed to be in their drinking habits. Our macaw, a lady at least in this respect, liked an occasional sip of sherry or wine. The Carcross macaw preferred a daily ration of rum.

The Klukshu Indian children were right. It would indeed be fun if Victoria's three prestigious parrots could ever get together. Louis, Polly and our Scarlett would sure have tales to tell!

One of the reasons we didn't take Scarlett to Carcross on our first trip was the weather. For six weeks that summer it rarely stopped raining and for several days it snowed.

One night when the temperature dropped below freezing we found a prospector's cabin that obviously had not been used for a long time. Little was left of the walls or windows and the three mattresses were completely covered in swallow droppings. The boys and I preferred to sleep under plastic on the ground or in the tent, but we put Scarlett inside the cabin beside the Coleman lantern. Despite the cold, next morning she was as objectionable as ever.

Considering she came from the tropics, Scarlett was surprisingly hardy in cool climates. She had survived the previous record-breaking winter in Victoria and had taken a trip with David to northern British Columbia where weather conditions were much more extreme. She did have an advantage in her method of travel: she supplemented her down feathers with the goose down in Dave's jacket. There were many double takes when, amid the ice and snow of 60 below, people caught sight of Scarlett's red-and-white wrinkled face peering out from his parka.

Yukoners were even more surprised to see a tropical macaw enthroned in their mountains atop a truck and rubber boat and sweeping her long red tail feathers across the summer snow. Even long association with the legendary Polly I of the Caribou Hotel in Carcross had not prepared them for that.

Further north in Alaska the weather got worse but we took refuge in a dilapidated, much-patched Quonset hut. From the outside it was a conglomeration of pieces and patches but it was sheer paradise inside. I nicknamed it our Paradise of Patches. Originally it had been erected by the government as an emergency shelter for stranded travellers. Recently it had been dragged down out of sight of the road to be used by fall hunters from Haines.

As we moved in, a family of Arctic ground squirrels

Home in the Yukon–our Paradise of Patches.

moved out. They returned each day while we were away hiking the tundra to ransack our supplies and exchange chatter with Scarlett. We found their company a constant delight and fed them foliage, flowers and peanut butter sandwiches–squirrel brand, of course! And when the time came to leave, we took a few south with us to study their hibernation patterns.

Porcupines were also common animals along the road. Unlike the ground squirrels, they had few friends among the sparse human population. The highway crew complained that porcupines were mutilating the road signs by eating the wood, and they shot them whenever possible. We saved one from such an ending by herding it in a cardboard box. After filming a fat, waddling ball of fur rapidly transforming into an outraged, bristling steel-hard tank, David decided to take it back to Island View. We called it Windi because the sign it had been nibbling had once said "Winding Road."

Now Scarlett had several companions in the cabin to keep her occupied while we filmed the abundant wildlife along the Haines Road. Her next room-mate was a gyrfalcon.

We had driven past the bird at fifty miles an hour before we realized what it was. An Arctic gyrfalcon, the largest of the falcons, was feeding on a roadside kill. We backed up slowly to get a better view. To our amazement the falcon did not fly but tolerated our approach to within fifteen feet, all the time continuing to pull slowly at the ground squirrel carcass.

"That bird is either extremely tame, sick, or so young it doesn't know any better," David said. "Let's see how close I can get to film it while it's still on top of the squirrel."

As David slowly approached with camera in hand, the falcon flew off to a nearby perch but with obvious reluctance. Dave then arranged a few nylon nooses around the bird's prey. Even before he could climb back into the truck, the gyrfalcon had returned to the ground squirrel and was caught in the noose by one toe. Quickly Dave disentangled the weak bird; surely the easiest capture of any bird of prey.

"It's a recent nestling and so thin and emaciated I doubt it'll live long especially with that peculiar behaviour. Animals so different from the norm do not survive in the world. But if this one lives, we can add it to the breeding collection," commented Dave as he carefully wrapped the gyrfalcon in a towel for a safe trip back in the truck.

We named the gyr Jerkin and arranged a comfortable perch for her at the cabin on an upturned mattress. Grant, our amiable, diligent and long-suffering assistant, surrendered his new leather gloves so that David could cut two long thin strips from the fingers for use as soft jesses around the bird's legs to restrain it to the mattress. Over the next few days Jerkin responded miraculously to being hand-fed ground squirrels and ptarmigan from an abundant supply of road kills.

A macaw, as well as the proverbial cat, must enjoy nine

lives. One day we opened the cabin door to find Scarlett sitting at Jerkin's feet.

"My God! She could be Jerkin's next meal," cried David in horror as he rushed to the rescue.

The macaw had a tiny hole and several scratches on her white cheek patch but was otherwise unhurt.

"Scarlett's lucky she escaped with only a few blemishes. No wonder Jerkin wasn't making much of a living in the wild. On the other hand, maybe the bird was paralyzed with surprise to find a prey that talked back. Perhaps Scarlett talked him out of eating her," commented David with a laugh.

Scarlett's habitual perch was on the clothesline which was attached to nails on opposite walls and strung across the room above the oil barrel stove. She must have abandoned her habitual perch by the door, beak-footed down a pair of jeans and waddled across the floor to greet the newcomer on the mattress, probably with an affable "Hello and how are you?"

No sooner had we hung Scarlett back on the clothesline than we smelled burning feathers. I spun around to see Scarlett's long red tail sizzling over the stove. Once again the fearless macaw had left her perch and picked her way along the clothesline. As if to demonstrate her method, she dangled over the frying pan, this time from a pair of socks.

To cut off her travel to dangerous places via the clothesline, we altered the placement of the line and built her a box over the door. It took her just three days to bite her perch to splinters. "What's up, what's up?" she exclaimed in astonishment, as having eradicated her support, she fluttered gracefully to the floor. We then had to build her a new box almost every day.

Later when Scarlett started to chew on the already much-patched walls of the cabin, we had no alternative but to take her with us during our forays over the tundra. If left alone she would be burned, eaten or else she would eat the cabin.

10
A Tortoise at the Typewriter

Somehow Scarlett survived our summer in the Yukon and Alaska, as did our assistant. When we all returned to Island View in the fall, Grant stayed at the beach to help build outside enclosures for Windi and the ground squirrels, and an inside one for Gypsy.

Gypsy's pen in the living room spanned half of one wall and extended from floor to ceiling. If she wanted warmth she could move to one end beside the pot-bellied stove. If she wanted to tease Scarlett she could exchange squeals and screeches at the other end. Only the length of the curtains separated her from the macaw's perch.

Now that the gibbon and macaw lived in the same room during the day, they imitated each other's vocabulary. Gypsy could not imitate Scarlett's words but she did pick up her other vocalizations. Sometimes from another room I wasn't sure which of the two was speaking.

The gibbon adapted to the new pen better than I expected, probably because now we were all penned together in the living room and she could always keep me in sight. Whenever I needed to put her in the pen, I lay her down on the ledge and pretended to change her diapers. Just like any baby, she conveniently raised her legs in the air and I would choose that moment to sneak away and close the door. As long as she could see me she was quiet but if I left the room and she was still in the pen she screamed. On occasion during the day I let her out and she happily swung on the curtains and pulled Scarlett's tail to the accompaniment of a harsh grating *Ouch!* When she learned to make long flying leaps to land in a box of correspondence or freshly typed manuscripts, I edged my desk further away and tied all the lamps to the walls. I waited with resignation for the time she would progress to the gleaming white incandescent lights above my desk. To an ape they must have looked like handy exercise bars.

Gypsy, Scarlett and I were alone in the house at Island View. Grant had gone home to Duncan to help his father build a boat. David went off to Oklahoma to narrate Al Oeming's film on how to catch Dall sheep for breeding in zoos. This was rather an ironic assignment as we had spent six weeks that summer trying, unsuccessfully, to catch one ourselves. No sooner did Dave return from the States than he was away again, this time to the Prairies to narrate our own film.

He left me with the company of the animals and with strict instructions to Write A Book. For years I had been thinking about it. In fact, I started one when I was eight years old but like many childhood projects it lapsed and became dusty with time. Rather than devise exciting adventures for fictional characters, I preferred to have the adventures myself. Like many Australian girls, I set out to see the world. For ten years I met characters I could never have concocted myself and now it was time to put them into print.

School resumed for the fall but this time without me. The phone rang constantly to lure me back to the classroom. "Sorry," I said before I left the receiver dangling, "I am Writing A Book."

But with tortoises at my toes, a macaw intruding her own

ideas or imitating the staccato hum of the machine, and a gibbon picking out the keys instead of me, there were too many distractions to compose a beginning. After an unsuccessful week trying to compete with my animal companions, Mother came over to remove them to her place.

I was alone. What were my family doing in Australia? I picked up a letter and looked at the date. Just over two months to Christmas! I was about to miss the sea mail. Writing A Book would have to wait till I had bought the gifts and written the cards. Do Not Open Till Christmas was a catchword of the season that seemed to typify my procrastination, and it was indeed Christmas before I got down to work and opened to the first chapter.

Just before Christmas David phoned from Saskatchewan and told me to take the train to Lake Louise. He wanted to read what I had written. As I hadn't yet started, I closed myself into the unaccustomed luxury of a roomette on the train, lowered the shutters on a superb panorama of snow-laden trees and frozen streams and started grimly on page one. The Rocky Mountains would have to pass me by!

At that particular point in my life my characters in print were having a far better time than I was. I consoled myself with looking back at the diary of my first fall at home and realized that delay was justified. Living with the animals had scarcely allowed me the time to Write A Book.

Scarlett, for the first time enjoying company in the house all day, would rehearse her whole repertoire of words for as long as I was there to listen. "Hi! How are you? What's up? What's the matter? Hi there! Yoo hoo! Polly want a cracker? Say something." And then when I did, she'd answer, "Shut up. Stop that."

And yet when I took her with me to a dental convention she wouldn't talk at all. Not even when she sat on the back of a chair at the head table of a banquet in the ballroom of the Empress Hotel and had six hundred dentists in her audience! While I gave a slide talk on what I hoped would be an apt topic–Teeth Of Animals I Have Known–Scarlett held a chicken leg daintily in her own leg and sipped wine offered liberally by her hosts, Dr. Screech and Dr. Freeze. As so much of Scarlett's language was scarcely fit for the polite company of the Empress, it was my good fortune that she did not choose to speak. Nevertheless, having no teeth herself, she served a useful purpose at the convention in illustrating what can happen when one eats too many sweets.

Gypsy, one year old and six pounds, was a constant source of entertainment. As soon as she heard my first of the morning murmurs and if I hadn't closed her door firmly the night before, she'd push it open, swing along the curtain rod, jump over the TV past the macaw, and leap onto my bed. Then in striking contrast to her boisterous aerial gymnastics, she would lean down very gently and kiss my face. Away again onto the bedroom curtains, over the dressing table, into the wardrobe up to the highest shelf, down to the pillows, she always returned for a reassuring kiss. By this time Scarlett would start talking for attention. It was impossible to lie in bed after that so I would get up and start the day. Although I was never hungry in the morning and neither was Gypsy, I used to sit down with her at the breakfast table and force both of us to eat something in preparation for the day's activity.

For Gypsy activity was endless. She literally flew through the house: into the kitchen to climb the shelves, into the bathroom to pull down the towels, into the bedroom to sweep off the covers, into the den to swing on the tripod and scatter the contents of a dozen tables. No movement was denied her. She ran and jumped and bounced and brachiated. She darted across her three-dimensional space trailing her elongated arms or using her long fingers to flick over bottles, boxes, cushions, anything that moved. She tried to eat soap and ash, to knock over pots and pans. Daily for our mutual safety, I learned new ways to gibbon-proof the house.

Gypsy's most ludicrous position was to hang upside down by her toes while she picked up everything within reach and pulled it apart. Her manipulation had improved to such an extent that she could spin the dial of my birth control pills, and of course everything discovered was transferred to her mouth. Although I often lost things with Gypsy in the house, I usually found them again–days later, in her mouth–erasers, bath plugs, rings, and like a lot of animals, sulphur tops off matches. She often disdained meat, fruit and vegetables, then within a few moments of refusing my offerings, she'd sprint off to steal lettuce from the tortoises.

If she was thirsty and couldn't find her bottle to feed herself (a behaviour which I discouraged because she habitually tore off the nipple with her teeth and spread the milk across the carpet), she opened the cupboard and took down the cardboard carton of powdered milk. Obviously she recognized the glass of foaming milk on the package. Other times she communicated her needs by putting my thumb in her mouth and sucking it.

On sunny days I took Gypsy outside with me while I did the chores. Now that she had an enclosure in the compound, I lay her on her favourite pillow in the pen and while she was quietly looking up at the ropes and the blue sky, I disentangled her clinging limbs from around my neck and skedaddled away. For a minute she wouldn't realize I had gone or perhaps she thought she was being changed. When she did see me depart she jumped over to the door, tried to lift it, squawked shrilly as if a leopard were after her, then squatted on the familiar pillow. Such a forlorn little figure clasping her long arms to herself and making me feel bad! I sighed as I remembered that her dependent period would probably last another five years.

Not that Gypsy was ever alone. The coatis next door usually put on a gymnastic display to attract her attention, and Pierretta the Canada goose spent much of the morning staring and honking through the wire.

On rainy days I left her in the living room and hardened myself to her cries as they followed me out to the compound.

During the afternoons while I typed inside the house Gypsy ran around the room, dangled from the curtain rod and kicked at the glass window or climbed up my back to watch my moving fingers. That was her least lethal activity and I enjoyed it till she squirmed around my neck and tapped on the keys with her feet as well as her fingers.

Towards evening her metabolism slowed down. Too tired for acrobatics but not yet tired enough for sleep, she glued herself to my neck, which made cooking my one daily meal very difficult. But it was while dining alone in the evenings that I most enjoyed her company. She would sit on my lap, play with my sweater, pick out minuscule pieces of fluff, comb my hair and groom me as her mother would groom her in the wild.

Sometimes at the end of a long day she acted like an overtired child. With mouth open, arms flailing wildly and uncontrollably, and making little *uh-uh-uh* gasping sounds, she jumped up and down on the spot, swept her hands in circles over the floor, fell over to one side, then picked herself up again. It was like the dance of some mad dervish. She collected an apple, ran with it to a corner, held it with two skinny fingers supported by a foot, dug into it with her sharp teeth, spat out the skin, dropped the rest, careered off under the chair, and dashed back to the apple. She chased it along the floor, took another bite, dropped it, and still chewing, finally flopped to a chair. With her eyes closed and losing coordination as if her mind were being blown by some drug, she staggered and stumbled around on the floor. Such behaviour would continue for an hour before she finally crept to my bed to sleep.

It seemed now that I was home during the day the phone and the doorbell never stopped ringing. There were the usual creatures requiring attention: a goose found stunned in a farmer's field, a saw whet owl with eye trouble, a red-tailed hawk accused of killing chickens, even a badly battered least weasel. There were the invitations to visit schools, to talk at clubs, to help stop an Amchitka blast or the rape of a forest in a provincial park. There were the door-to-door salesmen who stopped by accident at the end of our no-exit road then probably wished they had kept on turning. I'll always remember the Filter Queen Man. He was a little disconcerted to find an ape wrapping herself with his cords, a macaw screaming "Stop That!" to his every sentence, and then the final insult of all, when I kept calling his cleaner an Electrolux.

Some visitors came regularly. The pathologist came to take blood samples. Marilyn came with her mother to take pictures. Audrey came to help me pack anchovies for animals, a ton at a time.

Then Hilary came. As soon as I heard her knock I ran to the front door and we flung ourselves into the usual embrace. The welcome was exuberant and noisily unrestricted by worry about neighbours. It awoke Gypsy. She lifted the catch of her door and screaming with mouth open and teeth chattering, ran into the kitchen, chased poor Hilary around the stove, and when my harassed guest jumped on the table she lunged at her with great swipes of her arms. All the time and for an hour afterwards Gypsy made *OO-OO-OO* calls and low growls of a type and intensity I had never heard before. I recognized one as the loud, high-pitched alarm used, as Carpenter says, when a group of gibbons is surprised by a hunter or an enemy. The other was the deep-throated growl of a gibbon who is greatly annoyed and on the defensive.

Six months later when Gypsy was a year and a half old she demonstrated peculiar behaviour again when David came into the living room to kiss me. For the first time in her life she bit him savagely with little aggressive *huh-huh-huh* cries. Up till then she had just jumped up and down as if in severe anxiety, clinging tightly to my neck and trying to come physically between us.

It was only with Hilary and David that Gypsy was aggressive. Her attitude seemed to be one of protecting me from close "attack" as if she jealously fended me off the two people with whom I was most intimate. Gypsy still had some minor temper tantrums which were similar to those of a spoiled child: she opened her mouth and chattered her teeth, she dashed things aside and shook her long arms like a person who has suddenly burned her fingers. But only Hilary evoked her intense anger and rage.

One of the foremost authorities on apes, Robert Yerkes, has observed that captive gibbons who are strongly attached to their owners show their affection and sympathy by attempting to defend them against other humans. Several authorities have described instances where captive gibbons have resented the attention their human friends and companions have given to other animals or individuals of the same species. One even goes as far as saying that a gibbon may die if its human keeper shows a decided preference for another animal.

Researchers note that gibbon infants in the wild attack their fathers when they try to copulate with their mothers. Veterinarians observe that male dogs who are too attached to their female owners act like jealous lovers and snarl at their owners' boyfriends. Yerkes maintains that timidity, fear and terror are intimately related to resentment, anger and rage, and the former conditions may almost instantly give place to the latter.

In Gypsy's case I felt she was motivated by jealousy and a desire to protect me. She seemed to be defending me, her foster mother, against her foster father and another adult female aunt who had been in close association with her mother since the gibbon's earliest infancy.

Gypsy often demonstrated gentle, protective behaviour—to the coatis, to the white rabbit, to domestic kittens, and later to little cougars. Whenever she encountered an animal smaller than herself she would cuddle it in her arms and give it a kiss.

Gibbons in the wild have been seen to enfold another of their species in their arms or even, in captivity, a different species. The Hagensbeck Zoo in Hamburg described how a gibbon who already had a mate made friends with a stray cat. For weeks the gibbon and the cat caressed each other through the bars. Finally the cat forced its way into the cage, whereupon the gibbon threw her arms around the cat and refused to let it go. It was two days before the cat could get away. It did not return.

Sometimes such protection can be violent. Dr. Carpenter noted that in one captive colony, male gibbons joined together to attack any keeper who tried to catch any of the younger animals. This is unusual. One expects wild gibbons, who are normally loyal to their own small family, to be strongly antagonistic to members of the same sex. In the captive colony observed by Dr. Carpenter where male gibbons were housed together, the urge to protect the young of their enforced grouping was stronger than the urge to maintain distance between their rivals.

In such circumstances, a gibbon can bite savagely and inflict serious wounds with its keen canines, causing a clean cut which looks as if it has been done by a knife. Norm Winnick described how one pet gibbon of long years' standing suddenly turned on its mistress and inflicted such wounds with its canines that the lady had to be hospitalized and the gibbon sent to the Tacoma Zoo. According to Dr. Carpenter, there was a marked change in temperament in two of his captive females at the onset of their menstruation. Norm did not know if menstruation was a factor in the case he outlined, although the gibbon he received at the zoo had definitely reached maturity.

If handled by one person over a long period of time a gibbon can remain tame far into adulthood though it may be unpredictable with people other than its owner. The word "gentle" is used more often than any other to describe the temperament of pet gibbons. Older accounts maintain that gibbons are readily trained when young, are fond of petting and cuddling, and are appreciative of human kindness.

Tabraca, a Siamese missionary of the eighteenth century, even went as far as to say, in highly anthropomorphic terms, "One takes pleasure in raising this species of monkey because they are gentler than the others . . . friends of peace and full of compassion, they wish to embrace those who weep and their pity redoubles in proportion as they hear the groans of the unhappy; they never abandon those whose tears have not dried." I agreed with the priest wholeheartedly until I read: ". . . economical and dexterous, they never break or smash anything!" Obviously, he hadn't met Gypsy!

Gypsy was gentle except when Hilary came to Island View. And then she remembered her even though months might pass. It is generally agreed that gibbons have good memories. The *National Geographic Magazine* tells of a gibbon who had been raised for four years by an American family in Thailand. When they were forced to return to the U.S. without their pet, it was sold first to a dealer and then to the National Zoo in Washington, D.C., where the family, coincidentally, had gone to live. Quite by accident they saw a picture of their pet in the newspaper and hurried to the zoo to see it. After a separation of fifteen months the gibbon remembered its foster family and was as friendly and lovable to them as in Thailand, despite what could be regarded as a long period of rejection.

That fall the major interruption to Writing A Book was not Gypsy or Scarlett or visitors but my new job as film lecturer. One weekend Dave came home to recoup some of our losses in faraway places by presenting *Coast Safari* by popular demand for the second time in Victoria. Once the curtain had fallen on the last frame of the film, I relaxed and joined Dave on stage to acknowledge the spontaneous applause. Having done nothing myself during the performance except stand up at the back with Gypsy, I felt embarrassed but happy that the show was well received. This time as I stood beside him in the spotlight, my nonchalance was jolted by an unexpected announcement.

"And now, ladies and gentlemen, a special announcement. As you know, *Coast Safari* has been so well received that rather than disappoint those who have been turned away tonight, the management is pleased to announce an extra performance next Friday at 8:30 P.M."

"David!" I whispered desperately. "You can't. That's when you're going to the Prairies."

He smiled and continued with perfect composure. "But it will be *Coast Safari* with a difference. I have a commitment in Saskatoon myself. But as a special treat the film will be narrated by my wife, Lyn, and the other girl you met in the film, Hilary Stewart. So far you have heard only one interpretation of our adventures in the wilderness. Lately the girls have been making some backstage rumblings about other versions so I'm now giving them an opportunity to have their say."

I clutched the microphone in sheer horror, forgot where I was and stared at my husband as if transfixed. "Never! Don't be ridiculous. I couldn't do it. I bet Hilary doesn't even know about this..."–all of which resounded through the auditorium. Luckily any further reaction was drowned in an outburst of good-humoured laughter and another round of applause. I gave a sickly smile and stumbled off the stage.

In the privacy of the wings I turned on him. "Now what did you get me into that for? Half of the film is about your trip with Hilary to the Queen Charlottes and I haven't even been there. And how are you going to persuade Hilary to come over and do it? She probably has other plans for next weekend."

I sensed that further protests would be in vain. I knew fatalistically that I would return to that stage in exactly seven days to narrate the film.

I had to phone Hilary before she learned about her appearance in the newspaper. Her reaction was foreseeable. "David must be crazy! To spoil a wonderful film by having two inexperienced girls narrate after he has done it so well? You might be able to do it, Lyn. You've had the training and know about the animals. But why me? What do I know about sea lions or falcons? I can talk about Indian villages – but animals? No, I can't do it."

"Please, Hilary," I begged. "The extra showing has already been announced. If you don't do it with me, I'll have to narrate both halves myself and you can just imagine what a mess I'll make of totem poles. That's your specialty. Look, I'll spend this week typing out a script from a tape recording of one of David's performances and if you get stuck for words you can use that."

I really twisted her arm. But with my promise to send her a script by Friday morning, she agreed reluctantly to arrive in Victoria on the 5:00 P.M. ferry in time for the show. Usually standing up in public to address 850 people wouldn't worry me, but it makes a great difference when you know they have paid money to hear what you say. I meant to spend the entire week listening to the tape recorder and planning exactly what to say. Instead, I spent it promoting the advertising so there'd be 850 people to listen to what I was not preparing.

Every day I made the rounds of the newspaper offices, radio and television stations, arranging, advertising, holding interviews, phoning, postering and plugging.

As a result Gypsy and the coatis resumed the social contacts that had been severely curtailed by our weeks at home. They continued to be the terrors of television. On Pete's Place Gypsy was thrown out of the studio because she kept upstaging the coatimundis. So she flew upstairs to the control room and flung her arms around the producer. Perhaps she wanted to persuade him to put her back in the show. She obtained the opposite reaction, however, when she pulled his headset off and caused a temporary blackout.

It was Pixie and Pete that stole the limelight on the Jean Cannem Show. They scampered out of their carrying cases and made straight for a beautiful display of fruit and flowers that they must have thought had been specially prepared for them. Personally, I thought the coatis made excellent viewing for the television audience as they tore into grapes and latched onto bananas and tried their paws at flower arranging. Gypsy was superb as she picked out one dainty blossom at a time and held it up for a good look before taking a delicate bite.

It was obvious our interest in the display wasn't shared. The hostess, looking politely horrified, scooped up what she could from the very mouths of her guests and handed the remains to a technician. Then I understood her concern. The owner of the arrangement was sitting in front of the audience. As soon as the show ended, he whipped up what was left of his fruit and flowers, even the half-eaten apples, and left the studio without asking to meet the animals in person.

On the Ida Clarkson Show, Gypsy, not realizing she was the star and should be in front of the cameras, insisted on leaving the spotlight to hop on the backs of the cameramen and ride their cameras around the floor. Perhaps she was trying to encourage the coatis who were clambering all over the hostess' glamorous hairdo and recombing it with the sharp little claws I hadn't had time to clip. Pixie and Pete had been attracted to Jean Cannem's new hairdo in the same way. Both Ida and Jean coped marvellously with the coatis though I couldn't say the same about their coiffures.

I'm afraid I didn't show the aplomb of the professionals when the next day Gypsy once again lifted my wig. This time she chose to remove my veneer of glamour, not in the privacy of the classroom but in full view of the public at an important city reception. Two hundred strangers guffawed or politely tittered and I swore off womanly deception forever–at least in the company of apes.

Every night after the promotion activities I sat up past midnight with my ear on the tape recorder and my fingers on the typewriter. It was Thursday before I mailed Hilary's section and Friday before I typed mine.

Early Friday the theatre manager phoned me to say the 8:30 P.M. showing was sold out. He persuaded me to have a 6:00 P.M. showing to accommodate the overflow. I judged it discreet not to phone Hilary.

By five o'clock, dressed and nervous, I was at the ferry terminal with Sam the seal. Friends would look after Gypsy. Hilary greeted me with the words, "I had to go to Toronto this week and arrived back late this afternoon to find the script waiting. Sorry, but I didn't have time to read it over before the ferry ride. I'll never remember all those interesting details about the private life of a sea lion. Still, we've got three hours before the show. Let's have a nice, quiet, relaxing time so we'll do our best."

Hesitating just a moment, I plunged right in with, "Hilary, we're not going to have any nice, quiet, relaxing time. We're due on stage in half an hour. The manager scheduled a six o'clock showing."

"Lyn!" she exploded. "Why didn't you tell me? I could have worried doubly hard if I knew we were to do *two* shows."

"Well, I only found out this morning and I haven't yet read my script once," I tried to reassure her.

A little later, while a surprising number of people were pouring into the theatre for the extemporaneous performance, Hilary and I stood in the wings watching the minutes tick by to countdown. Neither of us had concerned ourselves with eating that day. All we thought about was having something to say, anything, while the film rolled relentlessly onwards, twenty-four frames per second for all of seven thousand two hundred seconds.

The lights dimmed. There was an expectant hush which I was expected to fill. I stepped up to the microphone. It was too low to pick up any words. In panic I forgot how to raise it. So in as lady-like a manner as possible, I bent down to remove my high-heeled shoes and started my introduction in stockinged feet.

I looked at the audience. Thanks to the spotlight on me, they were hidden in the darkness of the auditorium. After an initial stammer I launched into a surprisingly fluent, light-hearted eulogy of Life with Dave when really I intended to say just what I thought of him for getting me into such a situation. Come to think of it, considering he was so seldom around, perhaps that should have been Life Without Dave! Hilary was easy to introduce. She had been a boon ever since our early years with Sam in Vancouver and she remained an integral part of our pursuits when we moved to Victoria.

To an encouraging burst of applause and good-natured laughter from the unseen audience, I put my shoes back on and walked to the wings to bring her on stage. Instead of Hilary I bumped into the theatre manager. He was carrying two enormous bouquets signed, "To my two favourite girls. Love from Dave." That gave us the confidence we needed.

Not only did the audience enjoy the extemporaneous version of our adventures but we enjoyed it too. Our nervousness passed and gave way to hunger. The stage manager kept passing us chocolate bars between paragraphs.

After five hours of snacking on chocolate bars, the film finished for the second time and the audience gave us a gigantic ovation for which I was honestly embarrassed as the whole affair had been such surprising fun. Clutching our bouquets, we went off to a Chinese dinner with the theatre manager.

The year ended, as it began, in comparative catastrophe. Record rains and record winds sent the sea crashing around the house and into the compound. Gypsy, in her new orange Beaverlure snowsuit, was warmly tucked up at her grandmother's, but many of the other animals were almost inundated as the flood waters rose. Most of them survived without mishap. Some, like the seal, ducks, geese and seabirds, even enjoyed it. Pixie and Pete, turning cartwheels around their tightropes at the top of their enclosure, were high enough not to be affected. But a falcon and owl, like many in the wild, dived in pursuit of floating field mice and drowned. It was mopping up the mess and the mud after the flood that took the time and further delayed Writing A Book.

Well, there was always next year, I sighed, as we decided

to celebrate the cleanup at Sonya and Michael Gye's New Year's Eve party. David took Scarlett and I took Gypsy. As usual when she was in company, Scarlett sat on the back of a chair and talked little. But Gypsy, at home at her favourite baby-sitters', was the life of the party. She bounced on beds, leapt at windows, climbed curtains and played tag around our feet. She was poetry in motion. Only Scarlett was not impressed. Every time the macaw was looking the other way, the gibbon darted underneath the chair to pull Scarlett's tempting tail. Before Scarlett could croak a rebuttal, Gypsy flashed to the other side of the room.

The mischievous imp was more interested in action than eating. She ignored Sonya's New Year's Eve goodies. Although she couldn't be wooed by food, she could certainly be wooed by wine. The gibbon frequently quenched her thirst by going around to all the guests and sipping their drinks. With one arm curled around a handy shoulder, Gypsy would lean down and place her whole face within the rim of each wide goblet, gently sip or lick, then spring away to the next guest's glass.

At a quarter to midnight we were all startled by a loud noise outside. And in tumbled a dozen costumed New Year's Eve revellers. Immediately, and in contrast to her earlier indifference, Gypsy bounded to my shoulder, screeched a protest at the sounds of revelry and urinated. When I returned from the bathroom it was a minute to twelve. She faced the crowd of unfamiliar faces, dug her nails into my neck and screamed. If any entranced newcomer tried to pry away her clinging fingers, she bit him–so I was left standing alone and friendless at midnight while the crowd sang Old Lang Syne. Nobody wanted to kiss or hug me with an ape around my neck, certainly not Dave.

When the noise of celebration subsided, I zipped Gypsy into her fluffy new snowsuit, lay her down on Sonya's bed and soothed her to sleep. It was then a Happy New Year.

11
Cougars Come and Go

My New Year's Resolution to Get That Book Written crumpled under the crush of new arrivals at Island View.

When the phone rang this time, as it had been doing constantly all January, I expected another invitation to attack a local sewage referendum or support an anti-phosphate campaign.

"A friend of mine in Lake Cowichan has just..." I knew that this was not to be sewage or phosphates. With my heart beating in nervous anticipation, I listened while the caller tried to persuade me to provide homes for three orphaned cougars. It was the usual story of a hunter killing the mother then trying to sell the kittens. They would die if left in the wild. Zoos didn't want them as they had too many already. The hunter couldn't bring himself to kill them.

I did not want to be involved. It is too easy to lose your heart to a blue-eyed baby cougar and then after that everything becomes too difficult. I remembered the joys and the heartbreaks of the years with Tom and I did not want to live them all over again.

But it was inevitable. David who was in Portland when informed called the man back and accepted the cougars. Resignedly, I prepared the porch with papers and decided on the best formula.

In contrast to my cute cuddly kittens of other years, these new cougars were wild cantankerous spitfires, well adapted at three months of age to defend themselves against all who approached. Every time I came near they greeted me with flattened eyes and ears, open-mouthed snarls and well-developed fangs that crunched to the skin through two pairs of gloves. Knowing that the way to a young mammal's affection is often through suckling, I tried to feed them milk. But they ripped off the nipple and bit through the bottle. Thinking they were probably too old for suckling or were afraid of the imposed contact, I put the milk in a saucer. They tipped it over and padded through the mess. When I offered them meat, they bit my fingers instead.

Then they hid in corners and I lost them innumerable times a day. They squeezed under the chesterfield, scrunched behind the hot-water heater and crushed themselves between the wall and the stove. I gave them boxes for security but they distrusted mine and found incredible hiding places of their own. Trying to get them accustomed to my presence, I barricaded their escape routes with sheets of plywood but they knocked them down and continued to fly at me in their fear. They growled when they didn't know what it was they were growling at.

The big male was the most brash and in the forefront of the attacks. The other two were females and as yet indistinguishable in both their appearance and behaviour. They looked at each other for security and I knew none would be tamed while they were together.

I separated them, one in the porch, one in the living room and one in the bathroom. They whistled shrilly to maintain contact and tried leaping up the walls. They flung themselves against the glass. I boarded up the windows. I sat beside them, each in turn. They spat. It was impossible to sleep.

I tried taking the tom to bed. To my surprise he stayed

Orphaned cougars three months old.

there but I spent the night holding up the sheets so he could get air. At dawn he heard the calls of the others so I opened the door to unite them. Another sleepless night. I wondered how and when I would ever find the time and the space to write a book or even have a child of my own.

Finally I put them together in Gypsy's living-room pen. While they huddled on the lower shelf she brachiated above them, often pausing in her acrobatics to lean down and enclose them in her arms. But not even Gypsy's sweetness could mollify the orphans' reaction to losing their mother by a hunter who wanted her hide.

Hilary and Robin Bennett came over to help me tame them. We braved lashing claws and gnashing teeth by distracting them with gloved hands, seizing them by the scruff of the neck, wrapping them in blankets and stroking them behind the ears. We found we could make them go to sleep by passing our hands softly over their eyes. By avoiding any position of dominance above them and crawling on the floor at their level while they were eating or grooming or playing, we encouraged them to roll over. I plucked up courage and tickled their exposed tummies with my bare hands.

When Susan Im Baumgarten, an artist friend of mine, came to Island View to paint the cougars' portraits she was especially patient and spent hours trying to get them into amenable positions. We felt we might be making a breakthrough in the taming process when we heard one of the cougars purr. It was the tom and his animosity was forgotten in sleep. Perhaps his sisters would soon follow his lead.

When Hilary returned to Vancouver I went with her because we had donated our film to one of the secondary schools to raise money to commission an eagle carving. The students who hadn't done any publicity themselves asked me to do a few television shows to advertise their showing. Gypsy went to her grandmother's, as usual, and Hilary and I, with many feelings of foreboding, took the cougars to Hilary's apartment.

We kept the kittens in the bathroom where they immediately packed themselves behind the toilet bowl. Only very gingerly did we ever lower ourselves naked in their direction. As soon as Hilary's alarm rang to get her up for work, I dashed in to clean up the cougars' sand tray and feeding bowls and to rearrange the bathroom so that our hostess found some semblance of familiarity in her morning ablutions.

As the week ended, the cougars' confidence grew. No longer did they huddle in a heap when we entered the bathroom. They played in the bath and sat in the sink. As their interests grew higher we had to denude the medicine cabinet of its contents and string the towels near the ceiling. Hilary was wonderfully patient and even took pride in the fact that her apartment was probably the only one in Vancouver where it was possible to sit on the can and caress a cougar.

I was beginning to achieve some kind of rapport with one of the females. One day I chanced to take her to Jack Webster's morning open-line radio show. Though Jack has a heart of gold off the air, he has a national reputation for being blunt, brusque and outspoken. I wondered if his mood would match my recalcitrant cat. Then I remembered he ran sheep on Saltspring Island, an area of characteristic cougar animosity!

The area around Jack's radio station was impossible for finding a parking place so in desperation I drove on the pavement to the nearest gas station and pleaded for a space. A cougar in the bottom of a large wicker basket is handy for attracting service. The obliging attendant phoned another garage and I traded a lube and oil for space to park and a ride to the radio station. Declining the help of the friendly doorman in case he changed his mind if he found out whom he was helping, I carried the cougar upstairs and into the studio.

The timing was right. Jack, who had been labouring all morning on a monotonous topic, was delighted for a change of pace. He motioned me to a microphone and we had a gay romp on the air to choose a name for my cougar. The first three suggestions were Mercury Cougar, Van Cougar and Mitzi. We chose the last. As Mitzi seemed content to stay with Jack, I took advantage of the opportunity to run a few errands after the program was over. When I returned she was sleeping quietly at Jack's feet and I wondered if the soporific expression on the little lion's face had helped to soften the roar of the big lion. I didn't dare ask.

Now that the cougars were beginning to reveal divergent personalities, I felt ready to give them all names. The tom became Samson and the second female Chika. The only kitten that promised to be tractable was Mitzi. Alone they were all little monsters. Together they seemed to find comfort in closeness and were less likely to lash out their fear. I was surprised, as usually an animal becomes tamer if raised on its own.

I kept my other animal companions well away. Gypsy stayed at Mother's or in her living-room pen. Scarlett perched on the clothesline in the enclosed back porch. As soon as I let the cougars free in the living room they ran right across the room and dived under the chesterfield. I felt sure they would remain hidden in their newfound den. I sat on the floor in a corner and waited.

To my surprise, they peeped out from the skirting, padded around the room and advanced to the kitchen. I forced myself to stay seated as they knocked over the dishes, pulled down the towels and opened the cupboards. Ignoring my still figure on the floor, they passed me by and padded over to the bedroom. They followed each other along the dressing table and looked into the mirror. Startled by the strange cats of their reflections, they scrabbled in panic under the doilies in an attempt to get behind the images. Needless to say, the table-top contents were scattered afar and I could no longer remain rooted to the floor. As soon as I stood up they hissed and dashed under my bed, out of sight. And there I am sure they intended to stay.

By bedtime I was tired and frustrated enough to sweep them out with the broom. This only frightened them further and was retrogressive to our slowly growing friendship. I realized I couldn't keep them in the house any longer and that an improvement in our relations would have to come in the compound.

I put them in the food preparation room where I spent most of my mornings. They had one enclosure for outside exercise and a dark den next door for privacy. Although they learned to tolerate my presence and allowed me to touch them, it was never possible to enjoy the intimacy of an earlier litter I had befriended.

Gypsy was pleased to get back her mother's attention. But not for long. A week after Samson, Chika and Mitzi went to live outside, four more cougars came into the living room. This time the challenge was not in keeping them tame but in keeping them alive. They came from the Tacoma Zoo and like Gypsy were rejected at birth. Gypsy's mother had broken her infant's arm. The cougar kittens' mother chewed the tails off two of her offspring and the other two cats in her litter were taken away in case she tried to do the same thing to them.

I accepted the last two, a male and a female. I knew full well that once again cougars might break my heart, especially newly born ones.

The problem in trying to raise cougar kittens from the time they are one day old is their lack of resistance to disease, a natural immunity which they should have obtained from their mother's milk. Although Marilyn came out for the weekend to help me give the kittens around-the-clock care, the little female died.

We redoubled our efforts to save the male. Less than three pounds in weight, he was scarcely bigger than my hands. He seemed stronger than his sister and though his eyes were still tightly shut, he often climbed out of his cardboard box in response to my voice. While I tried to type he climbed on my lap, enjoying contact comfort in the same way Gypsy did. Marilyn named him Chico.

Determined to keep him alive, I took him almost daily to the vet's to check his condition. Although he waddled on oversized front feet, letting the back ones drag and flop behind him in the manner of all young uncoordinated cougar kittens, he ate well, put on weight regularly and appeared to thrive. Gypsy acted as Chico's baby-sitter when Marilyn and I were not available. She spent hours cuddling him in what seemed instinctive maternal behaviour.

By the time Chico was a month old he had started to make little stumbling forays around the house. Gypsy would follow and constantly enclose him in her arms. When Chico was a little bigger, over four pounds, he would try to disentangle himself from the gibbon's grasp and keep on walking. But with the ape's agility in all four limbs, the cougar could not get away. Gypsy was able to suck her thumb with one hand, pluck a flower with a second, snatch a blade of grass with a third, and clasp poor Chico with her fourth–all at the same time. Despite the unorthodox handling, the cougar enjoyed the company and the contact.

Chico, stay there while I change the lens.

Then just as Gypsy was freeing me from foster-mother duties by befriending the cougar kitten, another black bear cub arrived. This time Dr. Conway sent down Bubu, a roly-poly teddy bear whose mother had been killed by a logger. Once again animal orphans were to interrupt my writing schedule.

Bubu demanded more attention than the cougar kittens. She wailed her baby cries at the door, clawed the paintwork in an effort to get inside, and then when she did, encircled my legs with a bear hug or climbed up my jeans. Yet in contrast to Skookum and Cultus, she was much more amenable. She did share their habit of interspersing sucking on my arm and the bottle with attention-getting nips and bites.

Though Gypsy hadn't been jealous of Chico, she was of the bear. To prevent being strangled by both of them as they hung on my neck, I put them outside on the porch and gave them a big rubber ball. In the months to come, Bubu would be too big a companion for Gypsy but for a few short weeks both were able to burn off their excess energy by playing together. They were evenly matched in size. Gypsy's two advantages were in having four limbs to grab Bubu so the bear never knew which one was coming her way, and her rubber-ball bounceability to jump out of the bear's reach. Bubu's advantage was the strength of her arms and her stubborn tenacity to hold on grimly.

Most of the time Bubu was a happy little bear, galloping madly around the house and compound in a playful mood and almost smiling. When she got too rambunctious with either Gypsy or me, I cuffed her as her own mother would. Bubu was too boisterous to play with the cougar. Bears play hard but they also sleep hard. The gibbon and cougar had opportunities to play together during Bubu's frequent daily siestas.

At night Bubu and Gypsy slept together in the living-room pen. Gypsy was content to stay on the top shelf in the typical gibbon sleeping position, knees drawn up under her chin and girdled by her long arms. Bubu was supposed to stay on the bottom shelf. But being a ball of fire in between naps, she would teeter on her toes at the edge of her sleeping platform and stretch up to touch Gypsy's feet. The bear's long nose reached only far enough to nibble at the gibbon's toes. Gypsy's eyes remained tightly closed as if she was determined to ignore this playful persecution. Bubu tried more extreme measures. She pushed her nose over the edge of the shelf and bit. That did it! The gibbon leapt to a standing position and with teeth chattering, shot out her arms at the bear, as a snake darts its forked tongue. Startled, Bubu dropped back on her haunches to the safety of the lower shelf.

But she was a persevering little bear. Next she tried an assault from above. To get into a suitable position for attack, she had to cling to the wire with all four paws, leaving only her teeth as a weapon. Gypsy, hunched in the corner of the shelf, was safe and well outside the bear's reach. She bided her time till Bubu was enjoying one of her customary snoozes, then with three feet aflying, she dived at the unsuspecting bear. Bubu always woke up with a stupefied look but before she could attempt to retaliate, Gypsy was off and away in a flash.

Sometimes the bear and the gibbon were companions in crime. If their play around my typewriter became too riotous and unrestrained and they saw by my manner that I meant to impose a little discipline, they both ran to a hiding place behind the stove where they knew I couldn't reach them. I couldn't even see them except by climbing on top of the stove. As I peered down the crack between the stove and the counter, the two mischievous imps sat huddled beside each other on the floor staring up at me innocently as if to say, Well, here we are. Where do you want us to be?

Gypsy would be the first to creep into sight. She would peek around the edge of the stove, then make a mad dash to a second hiding place under the kitchen table. She knew full well that she was out of reach unless I decided to crawl on the floor and pull her out. By the time I did that, she would run into the bedroom under the bed. Hauling her out of that hidey hole took too much effort, and since she never stayed in one place too long, I always waited her out by the bedroom door. As she appeared I grabbed and deposited her on the top shelf. By this time Bubu would emerge from behind the stove. Less nimble than Gypsy, she was no match for my groping dive at the scruff of her neck. Once I made contact I held on grimly and installed her below Gypsy. At least one would go to sleep.

While all this activity circulated around my desk, Scarlett sat on her perch and made occasional comments on the action. She imitated the cougar's whistles, the gibbon's crescendo of *oo*'s and the bear's assorted cries and growls. Occasionally she was more specific. She told them they were "Bad birds" and instructed them to "Stop that!"

Once seduced by Scarlett's cajoling Spider-and-Fly whisper, "Come on, come on," Bubu climbed up the slippery sides of the TV set, clambered around the food tray and hugged her way up the tree to try and share the macaw's perch. I pulled her down to the floor, despite her painful protests. I knew Scarlett too well. As soon as she had lured Bubu up the tree with soft words of flattery and while the poor bear was clinging on all fours to keep balance on the swaying branches, she would probably change mood rapidly, let out a piercing scream and knock her to the floor with her beak.

But sometimes it was difficult to judge Scarlett's intentions. When Gal, a peregrine falcon, came in from the compound to spend a week in the living-room pen while convalescing from corns, Scarlett seemed to be in an amorous mood as if looking for a girlfriend. Showing an immediate interest, Scarlett descended from her perch, climbed across the curtains and stepped over to cling with both feet to the wire sides of the enclosure. She greeted a very bewildered falcon with her usual words of welcome, "Hallo, how are you?" Silence, not even a *kack-kack-kack*. "What's your name?" Still no reply. Undaunted, she spread her wings until they practically enveloped the outside of the cage, flummoxing the falcon even further. Finally she banged her beak against the door shrieking impatiently, "What's up, bird? Can't you talk?" I had difficulty in restraining my mirth. After another lengthy pause Scarlett muttered her parting shot, "Dumb bird!" I burst out laughing. Funnily enough, Scarlett joined in.

At first Gal hissed at the scarlet apparition then as she realized the wire protected her from assault, she calmed down, sat still on her burlap cushion and watched. Her beady eyes darted quickly back and forth between Scarlett's head and feet and tail.

Finding that her best conversation and display did not elicit a response, Scarlett tried cajolery as she next sauntered across the curtains. She moaned, imitated her kissing noises and kept repeating the phrase "I love you" which Hilary had recently taught her. When the falcon remained unmoved by such passion, Scarlett squatted on the window ledge to rest and cogitate on such an unusual situation. She spent an hour there alternately gazing at the falcon inside and the swallows outside under the window eaves.

The parent swallows darted back and forth in circles and

hopped on and off the clothesline. They dared not land on their nest under the guttering while the red predator was so close. To ensure that the swallows could maintain their daily worm intake, I had to intervene with the broomstick and transport Scarlett back to her perch on the TV set.

The next day Scarlett again found her way to the falcon and this time went through her entire repertoire of courting behaviour: fluffing out her feathers, squatting low on her breast, bobbing her head up and down, regurgitating her food and crooning to herself like a lovelorn Juliet.

Spring had arrived. The young swallows were already feathered. Wynken and Blynken had laid again. It struck me that in David's absence Scarlett might want Gal for a mate. On the other hand, perhaps Scarlett had an ulterior motive.

Day after day Scarlett continued to wander across to the falcon and spend hours in seemingly romantic overtures. Somehow she got her classification confused and persisted in calling the falcon a cuckoo. "Cuckoo, cuckoo," she would sing as she clasped the wire with her feet, rocked back and forth and regurgitated lumps of apple and sunflower seeds, despite the falcon's lack of appreciation.

When she started making a nest of torn-up newspapers at the falcon's feet, I decided to open the pen door and find out for sure whether Gal was a friend or a foe. I left the door open and waited, ready to run to the rescue. The macaw saw her chance immediately, clawed her way across the curtains, waddled into the pen and started to climb up the inside wire. The convalescent, a trifle quizzical but not perturbed, remained on her cushion on the ledge. "That's my girl," crooned Scarlett from the wire.

Then, with a lightning change of mood, her feet still clinging to the wire, Scarlett thrust out her beak like a rapier to attack the surprised falcon. There was a flurry of feathers, loud screeches from Scarlett and *kack-kack-kack* from Gal. I grabbed the double-dealing assailant in a flash before she could do any damage. "Ow, ouch," squawked the recalcitrant macaw as I returned her to the TV set before she could vent her spite on me.

Gal survived her stay inside the house and with her corns cured, she went back to live with her mate in the breeding pens.

But out in the compound there lurked a couple of bandits, the incorrigible coatimundis. One night Pixie and Pete, who had a natural ability as escape artists, chewed their way out of their pen and gorged on pigeon eggs. Instead of resting from their unexpected feast, the next morning they scooted over to the falcon enclosures. To lose the eggs of the prolific pigeons was of little concern. But to lose a peregrine falcon or even its egg would have been disastrous. It was traumatic enough that the falcons, who were about to breed, should be disturbed at all.

At that time the worldwide distribution of peregrine falcons had been much reduced and in many places they had become extinct. In Canada, although one subspecies was fairly abundant, two others were in a state of decline. Our peregrines were being held in captivity to encourage them to breed and hopefully release their offspring in the wild. Obviously Pixie and Pete could no longer have the freedom of the compound. They knew how to open doors, so Hilary came the next weekend to design an escape-proof enclosure with enough diversions to help distract them from further foraging.

It was just in time. I was still battling with Chapter One at the typewriter when Hilary came running into the living room with the dramatic news. "Lyn! The falcon has laid four eggs. It's Gal," she cried out breathlessly. "She attacked me when I reached in and tried to give her a chicken."

"Tremendous, Hilary!" I shouted in glee as I threw my arms around her and did a little dance, oblivious to the fact that Gypsy in the pen nearby might be incited to give poor Hilary her second attack for the day. "Perhaps Scarlett turned her thoughts to spring after all."

What's up, bird, can't you talk?

We both ran outside, tiptoed around to the back wall of Gal's pen and peeped through a tiny hole at the nest ledge. There in the gravelled depression by the darkest corner lay four blotchy red eggs, the very first ones hatched at Island View after several years of careful experimentation, and very nearly the first captive nesting success in the world.

The female had attacked Hilary to protect her nest site. Would she also protect her eggs and successfully raise her young if the eggs proved to be fertile? She had eaten her meal and was now carefully lowering herself over her eggs, gently rolling them into a comfortable position and fluffing out her feathers. Her mate, playing his more subordinate role, chittered in the background.

I knew that in other breeding experiments the first clutch was removed to an incubator, stimulating the female to lay a second and even a third clutch. This method allowed additional chances to produce fertile eggs and thus more young. As David was away again, I phoned Joe Simonyi, a friend who was engaged in similar breeding experiments, to help me remove the first clutch with the minimum disturbance to the parent birds.

As soon as it was dark Joe came around, heavily gauntleted, his overcoat collar turned up to protect his neck from attack. Feeling as guilty as a thief in the night, I stole along behind him as we sneaked into the falcons' pen. The male immediately flew down from his perch and latched onto my boot, tearing with his talons at my pant legs. The female, clucking in protest, lifted her great wings and hovered protectively above her eggs. While Joe crept around behind the female, I shone a light to distract her attention. Meanwhile the male continued to squawk and claw at my jeans. I pulled down my cap and tried to guard my eyes as the female came close to the edge of the ledge, her great grey talons lifted in attack. Joe gently picked up the eggs, transferred them to a warmed woollen tuque and moved quickly to the door. Bending down to protect myself from the falcon above and brushing the tiercel from my feet below, I followed.

Back in the house, we put three of the eggs into the incubator. Unfortunately, having been pushed to one side by the falcon and not incubated like the others, the fourth one was cold. But if all went according to plan, in two weeks' time the female would lay her second clutch and if these hatched, she had the chance of raising her own young.

Meanwhile under more controlled conditions, the first three eggs were safe in the incubator in the front room of the house. Or so I thought. Hilary returned to Vancouver and Gypsy regained her freedom to join the other animals in ensuring further delay in getting that book written.

It was May and fine enough weather for the animals to stay outside. The Arctic ground squirrels had come out of hibernation and were set up in a large cement-floored enclosure which Hilary had decorated with boulders and plants for a photogenic backdrop. I knew that squirrels could chew through wood but I hardly thought it possible for them to burrow through cement.

But one night the squirrels chewed a three-inch-wide hole in the cement and began tunnelling underneath the pen. The next day they piled up a mountain of stones, pebbles and rubble, later followed by sawdust which we had once laid under the cement to raise the level of the land in case of flooding. That night their busy little feet had dug down to the soil layer. Then the grass I had carefully arranged for photographs disappeared. The industrious animals had taken it all down their tunnel for nesting material. They could have easily continued to burrow their way out of the compound but they must have been satisfied with the accommodation I offered as every day they still came up for the fruits, nuts and seeds I left beside the pile of rubble.

Spring brought visitors again to picnic on the lawns by the front door. Scarlett resumed her audiences from her throne on top of the totem pole. Susan Sandeman-Allen came around to take Bubu and Gypsy for walks along the beach. Theoretically, this was supposed to free me for writing but the animals preferred to stay at home and get under my feet rather than go exploring with Susan. Once she got there, Bubu would enjoy scampering into the water, but it took effort to get her down to the beach. Gypsy had difficulty picking her way over the pebbled sand but she could run easily along the many logs and if attached to a human leg, felt confident enough to wade in the water. She even reached out to lick the salt. Nevertheless, if the animals had their choice, they all preferred to stay inside the house with me.

One day Susan managed to persuade Bubu and Gypsy to go for a walk on the beach. I closed the front door firmly and sat down at the typewriter with Chico and the best intentions to write Chapter Two. I was still waiting patiently for an adventure to come popping out from the blank page when a real one started within earshot. From the front room came a loud crash!

The incubator! The falcon eggs! Numb, I rushed out of the living room, through the kitchen and into the front room to find Gypsy dangling from a bar of fluorescent lights. The long fingers of one hand were tightly clasped around the narrow tube, her long legs were hanging in air and the fingers of her free hand gripped a peregrine falcon egg! David would banish Gypsy for life if she broke it. Maybe she'd eat it. I had to get that egg. The lid of the incubator had been knocked to the ground. One barn-owl egg, fortunately infertile, was a mess on the floor. The other eggs, by some amazing stroke of luck, were still intact. Quickly I replaced the lid to prevent a further decrease in temperature.

Thinking she had been very clever to discover an unlatched window, Gypsy *oo-oo-oo*-ed up the scale to attract my attention–hardly necessary at this point–and then leapt nimbly to the floor. Usually when I appeared she would spring onto my neck. This time she did not respond even to cajolery. As I hesitantly reached towards her, she bounced from the floor, bounded onto a counter and climbed up the shelves. All the time she clasped the precious egg in her hand. Astounding! She didn't drop it.

There might still be a chance. Not daring to hope and watching her from the corner of my eye, I pretended to walk away. Oh, how easily she could have dropped that egg! But she didn't. She climbed to the top shelf and placed it so very, very carefully between two cassowary feather head-dresses from New Guinea. With a whoop she jumped down to the counter, down again to the floor and then up onto my shoulders. Hurriedly I grabbed an apple from the kitchen and locked her in the living-room pen to eat it. I climbed up to the shelf, rescued the egg and returned it unharmed to the incubator.

That crisis over, the phone rang to initiate another which we didn't survive as well.

Chico, our personable little cougar, was a healthy four and a half pounds and thriving. Gypsy loved him and he reciprocated in the relationship by following her around on his bumbling big paws and sleeping for hours in her arms. He knew he had little to fear from the gentle gibbon but he snarled whenever boisterous Bubu approached. Chico was easing into my heart to help fill the void left by the death of

Honest, officer, I didn't do it.

Tom a year before. Chico also seemed to have all the incipient characteristics of a very special cougar.

Then his brother and sister arrived. For the past month they had been in the zoo hospital having their tails repaired. Cougars without long tails, one of their most distinctive features, are no use as zoo specimens so we had agreed to give the zoo some birds in exchange for the crippled cougars.

I knew there was something wrong as soon as I opened their box at the airport. It wasn't because their tails were bandaged or that they differed from Chico. Having been raised for a little longer on their mother's milk before she had rejected them, I knew that their white whiskers would be longer and their fur thicker. For particular aspects of their physical development, a human-devised substitute like Similac could never compare with the mother's natural milk. Yet my cougar Chico was more than a pound heavier, drank a whopping fifteen ounces a day and excreted solid feces. The new cougars, on the other hand, were thin and refused to drink at all.

As soon as I got home I phoned Norm at the zoo. He suggested a different nipple. I tried a dozen different nipples, but still the cats wouldn't eat. I phoned Dave in San Francisco.

"There must be something wrong with them," he warned. "Whatever you do, don't put them with Chico. Keep them separated." Guiltily I looked over to the guest room where all three cats had been lying together for the past hour. I flew from the phone, scooped up Chico's box and deposited it in my bedroom, leaving the others in the guest room.

I stayed up all night kneeling beside the new arrivals, trying different ways to get them to eat, but by morning they had only sucked a couple of ounces and had diarrhoea. I washed my hands, sterilized the bottles and took them to the vet. He could find nothing wrong with them but recommended intramuscular and subcutaneous injections of liquid protein and amino acids. He asked me why the mother rejected them, a question commonly raised in zoos and one that is seldom answered. Perhaps rejection in this case was not a syndrome of the mother's captivity but something in her offspring that she knew but we didn't.

That night Chico crawled from his box to my bed and slept on my pillow. By morning he regurgitated the little he had drunk the night before and his white excreta was curdled. I felt panic. I scrubbed my hands furiously before entering each room where the cougars were. I phoned the vet again. "Boil water" was his only suggestion. I always boiled water.

By 4:00 P.M. I realized that Chico would die. Suddenly his once-bright eyes were sunken, he lay on his side, panting heavily. David had come home that day and was busy in his den arguing the sewage referendum with a local councillor. I had to do something myself to help the cougar's situation, not just fill in time waiting for the end. The sun streamed through the windows. Was it too warm? Was it not warm enough?

I looked at Chico's hollowed eyes. He had stopped purring. I looked at his heaving body. He was not dead yet. Please God, *do* something. There was still time. I ran into David's den. One look at my tear-soaked face and the councillor stood up and walked to the door. "I'll leave you. This is not the time. I'll come again," he murmured. "Lyn, you have a lot of heart."

When we got to the bedroom Chico was dead. It was just too sudden. Yesterday he had been energetic, bright-eyed and playing in the park. My Chico. My Tom.

I ran outside, climbed the ladder and picked up the paintbrush, determined not to cry, determined never again to become emotionally involved with an animal. I had told myself that before. And I would tell myself that again. You must have the sad times as well as the happy times. You must have death to appreciate life. I remembered the clichéd advice but it didn't seem to help.

David called from the next room. "Let's inject another 100cc fluids into the other two," he said tersely.

Please God, let *one* live! We bent over the box. The two little cougar kittens lay prone at the bottom, panting like Chico. I knew further effort was hopeless. The autopsy would later report that "all organs were normal . . . no remarkable features . . . no cause of death could be determined."

I returned to the house. The litter box had been removed. The cardboard boxes were burned. The corpses were in the freezer to be picked up by the pathologist. The rooms were empty, silent. No whistles. No deep-throated rumbles of motorboat purrs.

I sat at the typewriter and looked at the blank page. But my mind was as empty as the rooms.

Silently the door swung open. With soft little *oo-oo-oo*'s, Gypsy crept to my lap, wound her arms around my neck and kissed away my tears.

12
A Sense of Humour Helps

It is well known that apes kept as pets demonstrate a strong desire to please their foster parents and are very sensitive to changes in human moods. Chimps, orangutans, gibbons, even the aloof gorillas, show tenderness and sympathy to other small, helpless mammals as well as to their own kind or their human companions. Vicki Hayes, probably the most famous chimp in literature, always responded with a hug and a kiss to her foster mother's distress. Gypsy, likewise, often changed her mood with lightning swiftness to respond to my tears with the same gentle actions a human might use.

More successful as an antidote to my occasional depression was Gypsy's ability to make me laugh. I don't think for a moment that she did this purposefully, in the same way she did when she communicated feelings of sympathy, but it showed me that things always have a funny side.

Like human toddlers, she was always getting into things and that usually spelled trouble. Like the time she leapt from the chesterfield to the swirl of sticky flypaper hanging down from the light bulb in the middle of the living room. Of course it gave way and she was well and truly attached to it by the time she dropped to the carpet. This was one situation where the gibbon habit of holding things down with all four limbs was most disadvantageous. Every fling she made with her arms or legs only fixed her more firmly to the long, tacky roll. It was enough to put her off flies forever. I ran to her rescue. Naturally, she clung to me for comfort and then the two of us were glued together in gum. We had to groom ourselves for days afterwards before we were finally free.

Gypsy was intelligent enough to know what was permissible and what was not, but like many children, if she could get her own way she did. She did not have the intelligence to comprehend my reactions when they were displaced in time from the event that had caused them. On such an occasion she would raid the kitchen sugar supplies at a time when my back was turned, then go and play unconcernedly in the living room. She always looked perplexed when I got mad after a quick look at the sugar bowl. Poor Gypsy! She had no way of seeing the telltale circle of sugar glistening around her flat black gibbon face.

It was easy to empathize with Gypsy. People cried with her and they laughed with her. Most often they laughed. She was a happy, lighthearted little imp and expressed in her movements, in her facial expressions, in her vocalizations, the ultimate in joyfulness.

The surest way to bring a smile to the face of an onlooker was to show him Gypsy at play, especially with a child. Scott Sandeman-Allen, Susan's youngest brother, was just the right age at eight to play with a 21-month-old gibbon. To the delight of visitors to Island View, the child and the ape played tag on the lawn, chased each other under the picnic tables, climbed the trees and swooshed down the slides.

But Gypsy liked best to play in cars, probably because she had spent the first year of her life travelling to school in one. Lots of cars were parked at Island View Beach in spring and summer and I was always on tenterhooks trying to dissuade the gibbon from playing around them. Many a beachgoer, after leaving his window open, was surprised to return to his car to find an ape in diapers prancing up and down on his seat and playing a strident tattoo on his horn.

One day I left Gypsy in my car while visiting Jill Bennett in hospital. As I sat by Jill's bedside a boy came running into the room and said excitedly, "Your gibbon has stolen my cap." It was true. I had left a narrow air space at the window through which the agile Gypsy had squeezed her fingers and lifted off the boy's headgear while he watched her at play.

I must be truthful. Not all were enthralled with my playful gibbon. There was the odd person with no sense of humour, like the man at the testing station. David had decreed on one of his fleeting trips back home that I was to take my ancient Austin through a safety inspection. I hate tests, especially mechanical ones.

"Why don't you take Gypsy with you to the testing station?" joked the postman after he ran my car around the block and listed some of its most pressing illnesses. "Perhaps the inspector will be more interested in her than the Austin."

"She goes everywhere with me anyway," I sighed as I packed the gibbon in the car with a new litter of orphaned cougar kittens that were being housed temporarily at Island View.

The reaction of the mechanic at the garage where I stopped to doctor the Austin's most obvious ailments was not an optimistic prelude.

"Your monkey is eating us out of house and home. She has just helped herself to my chocolate bars."

At times like these my ape really *is* a monkey. Gypsy knows exactly what she wants. Ignoring (fortunately) the cans of oil, the packets of car cleaner and bottles of brake fluid, she had climbed to the chocolates on the top shelf.

Back at the car she sat in the box with the cougars. One long arm was entwined around a kitten, another was ransacking my purse, and a foot was occupied in transferring chocolates to her mouth.

"Gypsy!" At the stern voice of authority she did as she

always did when she knew she was in trouble. Thumb in mouth, she clung tightly to my neck for sympathy. V-e-r-y difficult to chastize her in such a situation!

After plugging down some dimes for her ransom, I drove to the testing station. The officious, white-overalled man who hurried up to ask for my registration card didn't seem the type to be distracted by Gypsy, at least not in the way I hoped. With a forced smile I got out of the car and handed him my card, at least what was left of it after Gypsy had scattered the contents of my purse. As he took away her recent plaything, she sat in the open car window and stared contemplatively at his back.

Suddenly deciding that her position would make a good photograph, I lined up the car and Gypsy in the frame of my camera while the attendant continued copying from the partly eaten registration slip. Before I had a chance to click the shutter button he spun around to say, "You can't take pictures here, madam, unless you have official permission. Put that camera away." Feeling like a Communist spy trying to take pictures of a military establishment I decided not to argue.

The tester, still stony-faced, finished his notes then started barking instructions. "Turn on the headlights. Brake lights. Tail lights. Left signal. Right signal."

While I flicked switches on command he jotted something on a card, placed it under the windshield wipers and ordered me to move on.

The gentleman who checked the tires and brakes was friendlier. "You can get out now. I take the car through this one. Better take your monkey with you."

Deciding not to jolt his complacency by telling him there was a box of cougars on the back seat, I cheerfully complied. Let sleeping cats lie!

Once out of the Austin, Gypsy jumped from my shoulders and skittered down the line of parked cars. By the time I had reached the cashier with my card of punched holes, the gibbon was returning with a lady in tow.

"This wouldn't be your monkey, would it?" she asked sweetly. "It's just been chewing up the tissues on my dashboard."

"Yes," I told her as Gypsy relinquished the lady's hand and took mine. "I think there's a good chance she is. Next to cars, her second love is paper–typing paper, toilet paper and tissues."

Although Gypsy would have loved to remain at the station to continue her own testing procedures, I distracted her attention to the cougars still sleeping peacefully in the car, and looked up inquiringly at the cashier.

"Fifty cents, please," she said curtly, not even giving a glance to her unusual customers in the back seat. "Come back again when you've fixed up your head lamps, tail lamps, plate lamps . . .

The lady's list of instructions remained unfinished. Gypsy, bored with baby-sitting those that refused to wake up and play, had just noticed my glaring orange rejection slip and ever alert to something new, she leapt like lightning to the window, snatched it from the cashier's hands and tore it apart with her teeth. I laughed.

Gypsy, who hated to be rejected herself, dramatized my own feelings exactly.

13
A Star off-Broadway

Gypsy's good humour and playful antics may not have been an advantage in cutting the ice around the testing station but it certainly helped to add hilarity to film tours.

Dave came home just long enough to pack the two of us off around the province with our film *Coast Safari*. Film touring gave Gypsy plenty of opportunity to be the centre of attention and to indulge her passion for cars. Not that we always travelled in a car. We also went by bus, taxi, launch, crummy, logging truck, barge, tug, skidoo, toboggan, big plane, little plane, and sometimes even on foot. As it was rather awkward to take along a typewriter with a film projector, two big reels of film, a slide talk and a change of clothes, not to mention Gypsy, my resolution to Get That Book Written faded for another year.

Life at home had pitfalls. Life on the road had more, especially for a one-woman, one-ape, peripatetic projectionist team. Putting on a film tour sounds simple in theory: booking halls, buying ads, showing the movie. In practice it can be frustrating: janitors may forget the day of the performance and you have to break into the hall with the help of the local constabulary; the screen your host promised may be only handkerchief-sized and you have to run into the nearest house for a white double bedsheet; sometimes you borrow a projector to find that it doesn't focus or it breaks apart; other times you borrow a projectionist to find that he can't focus and the film breaks apart, winding in piles at your mesmerized feet. It's times like these you are glad to have a gibbon along.

Gypsy was accustomed to film tours. Mother had taken her to the Kootenays in the eastern part of the province where she charmed the audiences though, I must admit, not necessarily the motel owners. Fascinated by light bulbs, perhaps because they looked like fruit hanging from trees, she leapt to the lights in the middle of a few motel rooms, and bulbs became a normal addition to Mother's bills.

But Gypsy was a help far more than a hindrance. Once she did a whole show by herself. We arrived in Nanaimo on a day trip from home to find that one of the students had taken the adapter for the microphone home for the weekend. The janitor searched for alternative equipment but none had been found fifteen minutes before curtain time. While somebody went downtown to rent a sound system, Gypsy and I went on stage to entertain an audience who had been waiting since noon for the two o'clock show.

To demonstrate how attached a young gibbon is to its mother, I gently unwound myself from Gypsy's grip, put her down on the floor and walked away. Instead of running to me with her little *oo-oo-oo*'s and clinging to my leg for her usual ride around the stage, she decided to explore on her own. She dived right over the apron, scooted down the aisle, dashed back up the stairs, and completely ignoring me ran behind the curtains. The audience laughed but I lost some of my credibility as an expert in predicting the behaviour of apes.

For the next half hour she shinnied up and down the curtains, brachiated from the railings and demonstrated how gibbons use gym equipment. Once Gypsy leapt to the projection table, grabbed her milk bottle and treated herself to refreshments. The crowd thought it hilarious. Even if the audience had wanted to hear my learned commentary on gibbons, wild and otherwise, nobody could hear me above the sounds of obvious appreciation.

At last a truck with the rented sound equipment drove up to the hall entrance. As I walked off stage with Gypsy, there was an enthusiastic round of applause. However, to my horror, the men dumped the boxes on the floor at the back of the hall and walked out the door. "Sorry, forgot the adapter. Don't worry. Be back in a while," was the message relayed to me from a member of the audience before I could get to the truck.

All I could do was to return with Gypsy to the stage and put on another extemporaneous performance. Half an hour dragged by. I had just finished her life story, the story of gibbons generally, and was just starting on my own when the crew came back with the adapter.

It was now 3:30 P.M. and people were starting to arrive for the second performance. Impatiently I set up the equipment, put Gypsy away in the large projector box and turned on the switches. I launched at last into my official introduction. No sound. I didn't wait for the crew's excuses. In desperation I decided to delay no longer. Hoping the audience could hear me without a microphone, I shouted my way to the back of the hall. Fortunately Gypsy had helped to put the people in a sympathetic mood and they even murmured empathetically when I muffed the climax by getting to the end of the narration too late while the screen was flashing popcorn and candy time.

Occasionally a girl friend joined us on film tours. Audrey travelled in the camper with us when we did our interior trip to the Okanagan. As this time all the arrangements had been made in advance, we hoped to enjoy a few relaxing activities en route.

At the high school in Merritt the students watched spellbound as Gypsy performed for them on the lawn in front of the school steps. She chased their fleeing feet, pivoted on their arms, leapt after their coats, diverted an instant to whirl around the bars, then bounced back into the fray.

In Manning Provincial Park Gypsy interrupted a workman who was trying to paint a signboard. Having had prior experience with paint at Island View, Gypsy was very cautious as she leapt past a freshly painted support to brachiate from the unpainted brackets. While the painter stood scratching his head, Gypsy dangled by one limb and rhythmically kicked at his painting arm with her other three. Every now and then she jumped down and waddled over to touch me as I sat on the steps of the camper. My cocky little gibbon in plastic diapers hobbled bandy-legged over the gravel exactly like a human with bare, sore feet. The painter, obviously perplexed, wasn't sure if Gypsy was animal or human. Sometimes I wondered as well.

While Audrey and I set up camp in the park, Gypsy ran along logs and picnic tables. Although there were trees in abundance, she never climbed one. The only sign of her arboreal heritage was when she ran around with one hand held high above her head as if ready to swing if she wanted to.

At a home in Kamloops she chased Haida, a Vizsla pup who had once lived with us at Island View. Although she sped to catch the dog's tail then nimbly averted the playful repercussions, the gibbon was a little more defensive than in playing with the children. She was easily distracted and every half minute or so she interrupted her game of tag to touch or kiss our hostess' baby. Perhaps in playing with strange animal companions she appreciated occasional reassurance from a human.

At Boston Bar our tire blew. Jumping up and down beside us on the highway, Gypsy attracted so much attention that several people stopped to help us.

Gypsy enjoyed people. She was happiest when playing to

audiences in schools and theatres, parks and campgrounds. There were very few times she appeared apprehensive. She didn't particularly care for animals larger than herself. When Audrey and I walked around the Okanagan Game Farm, Gypsy clung tightly to me and ignored the animals just as she had ignored gibbons in zoos nearer home. She stayed close to me while we were near water. Camping chores were a little cumbersome with Gypsy slung around my neck when I knelt by streams. I kept getting her foot in my mouth instead of the toothpaste, and a couple of her toes in the dishes instead of soap.

On the Vancouver Island tour Gypsy and I took Sue Sandeman-Allen with us. Or rather Sue took us. With my mood sometimes as flighty and excitable as a gibbon's, it was good to have as our companion the calm, capable, practical Sue. Although a student in grade twelve, she had taken off time from school to go with us; but such is the friendliness and flexibility of people in more remote places, that Sue probably learned more on the road than had she stayed in the classroom. Especially in the company of a well-schooled gibbon.

Although David had blithely informed me before our departure, "Don't worry about the logging roads, they're better than superhighways," our first one remains in my memory as the Nimpkish Nightmare.

It runs through the Nimpkish Valley from Gold River on the west coast of Vancouver Island to Port McNeill on the east. The guide books say it is beautiful. Sue and I wouldn't know as we had to drive it at night. "Night is the only time the general public can travel the Nimpkish Road," David had said. "It will be as busy as a freeway so you'll have lots of company. If you break down, remember you've got two spare tires and a jack and there's sure to be someone to help two maidens in distress. If you leave Gold River at 11:00 P.M. after the film showing, you'll be in Port McNeill well before 2:00 A.M."

We left Gold River before 11:00 P.M. all right but with slightly different words of advice ringing in our ears: "Drive carefully" . . . "Avoid potholes" . . . "Concentrate on keeping on the right road. There are lots of side roads and Y branches to fool you." And it was 5:30 A.M. the next day before we arrived in the civilization of Port McNeill.

With all that advice, none helped our main problem of how to keep awake. The day before, like all days on a film tour, was a busy one: driving for many miles along narrow, winding mountain roads, touring the modern instant logging towns, lecturing in the exciting new school, showing Gypsy to her admirers in the shopping plaza, having dinner with the staff and teachers and finally putting on the film.

Gypsy was sound asleep in the camper. Sue and I longed to join her. The first twelve miles were supposed to be the best. We found them the worst. The narrow one-lane road wound high above the river valley and I was a little disturbed to see our truck was level with the trees. Unused to driving mountain roads at night, I cautiously kept the speedometer to a maximum of twenty miles per hour. To keep me awake Sue fed me our dinner dessert of sticky cream pie and oranges, one piece at a time. When eating had no effect, I sang, whistled and pulled my hair. Finally I gave in and Sue drove. Rather than snuggle beside Gypsy and go to sleep, I found that my fear of Sue going off the road kept me awake. When my long-suffering companion topped 15 mph I accused her of speeding.

At midnight we stopped at Oktwanch Gate where a lady in a trailer gave me a permit slip to fill in. I fondly imagined it was recognition of our presence on the road and that if we were still loafing along in the morning when the logging trucks came zooming by, she would suspend production till we were cleared out of the area. In actual fact when later I read the small print, I found I had promised to drive the road at my own risk. One of my other promises was not to "discard any bottles, cans or other refuse." This was ironic as one of the bits of advice we had been given in Gold River was "Avoid both cars and cans."

The middle section of the logging road was wider, the terrain more level and there were less potholes. We were nipping along at about 25 mph when at 3:00 A.M. the tire blew!

Practical Sue jumped out of the truck and started to work the jack. No longer lulled by the rhythm of the engine, Gypsy awoke and started screeching a protest. While I attended to her complaints, Sue added some of her own. "Lyn, I can't get this jack to work. It doesn't fit."

"Well, according to Dave, this superhighway is supposed to be alive at night with lights. We may as well wait for somebody to stop and rescue us," I suggested philosophically, thinking mainly of snatching a few moments of sleep.

Never dreaming for a moment that a car wouldn't stop, we changed our best dresses for jeans and warm sweaters and dozed in our sleeping bags. At 4:00 A.M. a vehicle sped by but by the time we had stumbled out of our sleeping bags, it was only dust in the distance.

"Perhaps if I blow the horn he'll come back," I said desperately, jumping in the driver's seat. I persisted in blowing a series of long and short blasts in what I thought was an SOS signal. "Lyn," Sue laughed, "that's not SOS. It's OSO." It didn't matter that I got my signals crossed. The vehicle did not return. Not daring to go to sleep again, we jogged up and down the road to keep awake and wished we were carefree gibbons.

At five o'clock another car came along and we waved wildly to attract the driver's attention. He saw our problem and using his own jack changed our tire. His parting words were, "Tell your husband to put his jack back in his Mini Minor!"

Three hours later we passed through Kokish and Beaver Cove and cleaned up at the gas station in Port McNeill where it was gently suggested that I ditch the blown tire. At eight o'clock we changed back to dresses and drove to the school for the day's lectures and film shows. It was a relief to leave behind logging roads for a while and sing along the black top to Port Hardy.

Gypsy's favourite place in Port Hardy was not the school but the newspaper office. Having taken her out of many glue pots and wiped her clean of much printer's ink, I did not turn her loose until Neville Shanks, the owner of the *North Island Gazette,* strung a rope across the entire room and asked to take pictures. Brachiating with abandon, Gypsy stayed in mid-air, and the ad layout stayed safe beneath her dancing feet.

Another couple of animal nuts in Port Hardy were Graeme and Andrea Conway, who had a pet white poodle, Cory. After the film show we all returned to their house for our own entertainment. While I guarded the Eskimo artifacts on the shelves and Sue guarded the lamps and glassware, Gypsy met Cory. Usually dogs, whatever their size, are wary of such strange-smelling animals as gibbons but Cory, like the Conways, was adventurous and friendly.

Approaching with the normal gibbon greeting, Gypsy wound her arms gently around Cory's neck and kissed her on the nose. Initially the dog stiffened. Then she took a chance and came forward, wagging her tail. From then on it was a merry chase, around the chesterfield, onto the chairs, behind the light stands (which were not tied to the walls as they were at Island View) and under the tables. When the going got too rough for her, Gypsy crawled into small spaces or leapt to my neck. By the smile on both animals' faces we knew they were enjoying themselves. Cory didn't even mind when Gypsy made a sneak attack at her tail.

The Conways had another pet, a ringtail monkey named Mandy, whom Gypsy completely ignored as she always did with animals closely related to her. Every time I took her to the zoo to introduce her to the gibbons and expected her to join in with their chorus of ear-splitting hoots, she dug her nails into my neck and averted her face. Her behaviour was the same when she saw Mandy. She relaxed only when the ringtail monkey was put in the porch.

Next day after driving another logging road, even more muddy and potholed than the Nimpkish Nightmare, we showed the film in Port Alice. In this modern town on Quatsino Sound the hall was packed and the people were even more informal and friendly than in previous communities. I shared a table on stage with some of the audience and sometimes even shared the narration. Instead of packing Gypsy away in the projector box because she was too much competition, I relaxed enough to let her sleep on a number of laps during the performance.

Back in Port McNeill the B.C. ferries agent, Marie Huggins, who was charmed by Gypsy swinging in her fruit trees, took it upon herself to make arrangements for an unscheduled showing of *Coast Safari* in her community while Sue and I went to Alert Bay and Sointula. We left our truck in Marie's back yard and, heavily laden with suitcase, projector, screen, sound equipment, cameras, reels of film, not to mention Gypsy, we dashed to get the noon water taxi.

Gypsy, always good for public relations, made sure we were the centre of attention. The friends she made helped to get our gear and ourselves off the dock. Once aboard, the gibbon skedaddled down to the cabin to find the passengers. She leapt from lap to lap until she found the one she wanted. It belonged to a tall, husky Finn. While the rest of us took seats on the side, Gypsy and the Finn played tag. They looked ludicrous, this blonde giant of a man and a little black imp, as they faced each other in the ring like two ill-matched boxers: jumping up and down, lunging in and out, darting backwards and forwards. Gypsy swung on his arm and kicked him in the face. He got her on the ground and bowled her into a somersault. The gibbon must have enjoyed it as she always came back for more.

Norm Martin, the school principal, was at the dock in Alert Bay to meet us and take us on a tour of Cormorant

Island. The cemetery in the centre of this Indian village and fishing community is lined with totem poles, each looking out to sea in the traditional manner. Gypsy was familiar with totem poles but with Scarlett monopolizing the roughly carved one at Island View, she had never yet climbed one. Now was her chance.

Chased by the usual knot of entranced children, she ran in and out of the totem poles till her pursuers got too close then she nipped to the top of one. In the cemetery, carved memorial poles marked each grave site. Gypsy by-passed all the poles that had animal motifs on them, like the killer whales, the ravens and the thunderbirds, and picked out the only one that was human. While we photographed she clung to the large face of a man colourfully painted in the modern manner and surmounted by a tall black hat. The Indian children looked up at Gypsy's diapered black body wound around the neck of the man on the pole and giggled. They were certainly fascinated by this unique visitor to Alert Bay, but I kept casting nervous glances at Norm busy at work with his camera. Our behaviour did seem a trifle unseemly for a cemetery.

However, the people of Alert Bay were friendly and didn't appear to mind their unusual visitors. Sue and I were invited to their homes after the film that evening and next day to St. Michael's Residential School where Indian children from many communities received lodging and schooling. Accustomed as she was to schools, Gypsy had great fun scampering around the classrooms and then at recess and lunch joining the children in another merry romp chasing from tree to tree on the spacious lawns. The boys were quick but the nimble imp was quicker. They thought it hilarious when she swung from the branches with one limb then used the other three to tug at their hair or kick at their faces.

Despite her ascent of totem poles in a cemetery and hand-adzed beams in a longhouse, Gypsy was an instant success in Alert Bay. It seemed that all the children in town came

Gypsy chooses a human totem in Alert Bay.

down to the dock to help us get our equipment into the boat for the ride across the strait to Sointula on Malcolm Island.

The people of Sointula were mostly Finns. Their friendliness and simple charm was typified by the motherly caretaker of the community hall who never having seen me before, threw her arms around me in an enthusiastic welcome as soon as I appeared at the door. Even Gypsy appreciated the warm welcome and jumped on the lady's shoulder with a gentle gibbon greeting, rather than what she might have done if it had been Hilary or even her Finnish friend on the ferry. The audience that night was the most responsive of the whole tour.

14
The Queen of the Charlottes

Back at Island View after a ferry ride south to Kelsey Bay and the truck down to Victoria, I walked into the kitchen to find a note on the table.

Dear Lyn,

I've had to go to Seattle. Just pop over to the Sechelt Peninsula and show the film, will you! It's for the Handicapped Children's Association.

Love, Dave.

So while Sue went back to school to write a report on logging and fishing communities and Mother continued to look after the animals, Gypsy and I set off up Island again.

From the Comox ferry we drove into Powell River, home of what has been reported as the largest single newsprint production plant in the world. The streets were deserted, the houses and the huge kraft mill were silent. Almost everybody was out of work because of an industrial dispute. Before Gypsy's arrival, the sole topic of conversation had been the strike and the atmosphere was grim. True to form, the lighthearted little ape, like some aerial Pied Piper, flitted everywhere, through the schools, along the streets, inside the shops, with a throng of children in pursuit. And the mood of the town changed.

The only thing Gypsy didn't like was our hike up a nearby hill to sit on a rocky knoll and look out over the sea and the islands. She squawked all the way to the top and down again. One of the children called out, "There's a big leopard behind you, Gypsy, you'd better watch out." To our astonishment and amid much laughter, Gypsy leapt with piercing *oo-oo-oo*'s to the safety of the nearest neck. Leopards are probably the only predator to kill gibbons, so it was an amazing coincidence. At such times the little ape could be uncannily human. As one of the teachers later remarked, "Well, we are not so far removed."

On our tour through the province Gypsy and I had travelled in a car or truck. Sometimes when we reached our destination we had to find alternative forms of transport. In Grassy Plains there were no ferries running after the film ended so we accepted a lift across frozen Francois Lake in a toboggan pulled by the ferryman's skidoo. With all the equipment clutched against my chest to safeguard it from the bumpy terrain and with our eyes shut tight against the bullet-like falling snow, it was an exciting and memorable moonlit ride.

But there was one place where we couldn't go by any of our previous methods of transport. To get to the Queen Charlotte Islands we had to fly. Although I had heard of one family who travelled all over the world by ship and plane with a pet gibbon who was never placed in a cage or travelling box, I didn't think Pacific Western Airlines would like me to book a seat for an ape, however civilized. And she would die of fright if left in a luggage compartment. So Gypsy went to the Queen Charlottes in a hatbox.

It wasn't a very modern hatbox, the kind often seen in the hands of fashion models. In fact it was a decrepit and old-fashioned hatbox, provided by the Goodwill secondhand store. Father had punched holes in the lid for windows, although they were soon enlarged into doors by teeth and feet. He kept it together miraculously by regularly adding lots of wire and black electrical tape. My parents-in-law had often used it to take Gypsy on outings, and when baby-sitting they put her in it at night rather than let her loose in the house. They looked it over, reinforced the lid where Gypsy had tried to use it as a trampoline and pronounced it perfect to fly to the Charlottes.

The day was grey and fog-patched as we drove to the airport. I surrendered my suitcase and film equipment to the conveyor belt and kept my hand luggage, consisting of Gypsy in her hatbox, on my lap for the short flight between Victoria and Vancouver.

While waiting in the lounge for the next flight to Sandspit, I put it on the chair beside me. A group of art students sketching the atmosphere of an airport kept looking at me surreptitiously. I wondered if they thought my hatbox contained some strange kind of hijacking equipment. I was tempted to open the lid and let Gypsy explode from the hatbox, but I restrained myself and looked circumspect.

"Your Pacific Western flight to Comox, Port Hardy and Sandspit is now ready for boarding."

As I strode across the tarmac, fashionably modern at that time in a mini skirt and high boots, Gypsy started twittering. Not wanting to take any chances until we were airborne, I whistled my way to the waiting Boeing 737 jetliner.

The noise of the engines warming up as we mounted the steps was deafening. Any cries Gypsy made would not be heard above those of the plane. Accustomed to the dark of the hatbox and thinking it was night, she remained silent and felt secure while we boarded the plane.

I sat in a double compartment so I could place the hatbox on the opposite seat. The stewardess came up immediately and volunteered to put it above my head in the rack. "No, thank you very much," I acknowledged sweetly. "I may

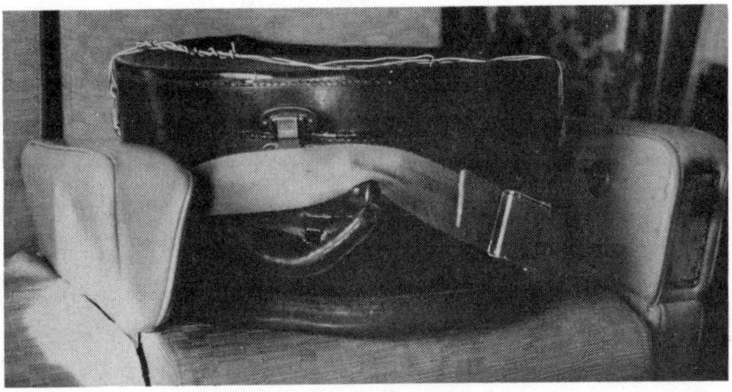

I'll keep it where I can see it, thanks. I may change hats during the flight.

want to change hats during the flight."

As soon as she left to attend to more appreciative passengers, I fastened a seatbelt around the hatbox, whipped out my camera and took a picture. Passengers in the nearest seats politely hid their curiosity behind their newspapers. I doubted if any could have guessed what was in the hatbox.

At takeoff I thought Gypsy might have wanted to take off herself and come flying through the roof of the hatbox, but accustomed to sleeping in it at Mother's, she stayed quiet. She did poke her black shiny eyes out one of the holes as if to ask, "Isn't it daytime yet?" As the pretty flight attendant brought along a welcome cup of coffee and beef-in-a-bun sandwich, I draped my coat over another hole through which Gypsy was now poking her long furry fingers.

"May I take your coat for you?" the hostess inquired politely.

"No, thank you very much. I may be needing it," I replied casually.

Feeling guilty at eating and drinking in front of Gypsy without offering her any similar refreshments, I covered up a few more eye holes in the hatbox.

Three hours later we touched down at Sandspit on Moresby Island. For a moment I thought I had landed in the

Caribbean. The community's park-like setting, wide sandy beaches, flat terrain, sparse trees and glorious sunshine looked more tropical than north temperate. My reverie was broken by a cheery voice calling, "Lyn Hancock, we've been waiting for you."

Neil and Betty Carey, adventurers, beachcombers, writers, were the most welcoming couple imaginable. Betty, with a wide grin and eyes that bubbled with life and enthusiasm, had once paddled a dugout canoe from Alaska to Seattle.

The Careys piled my stuff in their truck and drove us to their house. Crammed to the ceiling with bookshelves and with work desks overflowing with diaries, typing paper, pens, pencils and partly finished manuscripts, Gypsy and I felt right at home. Gypsy especially loved their yard. One fence was built wholly of driftwood, another was composed of whale vertebrae. Glassballs, skulls, floats, shells, bottles, remnants of ocean-going rafts–the list was endless and Gypsy found innumerable objects to whet her curiosity.

Too soon we had to leave and take the Misty Islands Transportation Company bus across the Skidegate Inlet to Skidegate and Queen Charlotte City. The rattling, yellow, dust-covered MITCO buses were almost as decrepit-looking

Listen, mister, strap-hangers should pay quarter fare.

as Gypsy's hatbox but the warmth and friendliness of the people who travelled them more than compensated for outside appearances. As we rolled along the shoreline bound for our launching pad on Graeme Island, Gypsy did the rounds of the bus passengers by kissing each of them delicately on the cheek. Formalities finished with, she then had a great time riding the stanchions of the bus, swinging across the aisle to pat the head or remove the glasses of the amiable bus driver, or bouncing from one seat to another from the back of the bus down to the front again. Everyone loved her and voted her the most riotous passenger ever to board a MITCO bus. The ape lost her aplomb a little later when we rolled and pitched over stormy seas in a tiny barge to Skidegate Landing. A week before, the barge had sunk. In common with the rest of her kind, Gypsy had always feared water despite her surprising tolerance of being bathed in a basin for her first few months. Not even her hiding hole buried at the bottom of my parka could ameliorate her terror or muffle her screams, and my skin bore the brunt of her long nails when we finished our ferry ride.

Once on dry land again she became her usual cheery self and helped me hitch a ride with Les Bullen, the Superintendent of Instruction. Our route from Skidegate to Queen Charlotte City was almost as bumpy as the barge crossing. But as we eased in and out of the muddy potholes along the winding road that skirted the rock cliffs, I could sense something of the captivation these remote and misty islands have for all who visit them.

At the Queen Charlotte Composite School every grade from one to twelve was assembled in the gym for our slide presentation. The children, mostly Indian, were fascinated by Gypsy's aerial acrobatics. During question time it was impossible to compete for attention when she was leaping from head to head around the gym. I finally gave up and we just sat, stared and enjoyed her.

Next morning we hiked through the forest so Gypsy could

climb the most famous tree in the Queen Charlotte Islands—the Golden Spruce. It is the only one of its kind known to exist. Its seedlings produce only green spruce yet the famous Golden Spruce of the Charlottes is burnished yellow.

We hitched to Masset in a truck. For some reason I couldn't fathom, Gypsy seemed petrified of the driver and preferred to huddle in her hatbox.

As the first half of *Coast Safari* is the rubber-boat trip around the southern Queen Charlotte Islands to visit abandoned Indian villages, I expected the film would be well received in the Indian village of Old Masset or Haida, as it is presently called. People began to arrive well over an hour before the film was due to start. I could foresee a problem fitting everybody in, and scheduled a second showing.

Gypsy responded to the excitement of the occasion by staying awake for both performances. I placed a long table in the middle of the hall and Gypsy used it as her personal stage while I tacked up bedsheets and dusted furniture. As the Indian children crowded around, Gypsy ran up and down the table, pushing heads, pulling hair and picking at sweaters. She played her favourite games of Pick and Pluck and Hit and Run. Yet when mothers came up with babies in their arms, she leaned down to greet them with a quiet kiss and a cuddle.

On this particular night, in addition to her usual job of providing the prefilm entertainment, Gypsy enlivened the film narration as well. While the story unfolded on the bedsheets, she had great fun leaping from head to head along the back rows, then, making use of every available hook or hole, walking along the wall to almost ceiling level. Her antics were far more engrossing than the film. And then all of a sudden we got both the film and Gypsy at the same time. She swung in front of the screen—and there, climbing the totem poles in the jungle of a Queen Charlotte Islands rain forest, was a lively black ape from Southeast Asia. The audience roared as her shadow continued to dance across their ancestral villages. I wondered again about being sacrilegious and pondered whether or not to pack her back in the hatbox.

Then Gypsy produced her *pièce de résistance* for the evening. One moment she disappeared. The next the whole room sprang suddenly into light. All heads turned towards the wall. Gypsy had flicked on the light switch.

Next day I tried to reduce her hyperactivity by a long hike through an actual rain forest near Juskatla. One of the local teachers told me that "hidden somewhere in the big timber far from the sea" was a giant cedar out of which the Haida Indians had begun to carve a huge sea-going canoe. Gypsy and I set off to find the unfinished canoe.

Crashing through the matted undergrowth and windfalls is very difficult for a creature as airborne as a gibbon. But she made no attempt to swing on the treetops in the manner of her free-roaming relatives. Unaccustomed to dense forest, she preferred to follow me on two feet in human fashion. But being only two feet high, hiking in the forest was difficult for her. If I got too far ahead she would send her shrill screams resounding through the greenery till I waited for her to catch up. She was adept at running across logs and swinging from branch to branch if they were low lying, but she never wanted to get too far away from me by climbing upwards. When she came to the end of one log and was faced with a leap into a large patch of salal, she seemed completely baffled. With my long legs I could stride through or over the thick undergrowth but for her to negotiate salal was like wallowing in a swamp.

Although a tropical animal noted for its delicate health in captivity, Gypsy seemed unaffected by the cool dampness of the Charlottes. She never shivered and was just as lively on the seashore or in a rain forest as she was in a heated building. Nevertheless when a light rain peppered the treetops, I continued the search for the canoe by letting her ride in my parka.

After stepping and stumbling over the stumps for the next

half hour, we came at last across the unfinished canoe. The Haida Indians built some of the largest dugout canoes known to man, some over seventy feet long and eight feet wide, canoes capable of voyages south to California. This one, lying where it was first carved at the foot of much less mighty trees, was probably more than eighty years old.

As I stopped to reflect, Gypsy leapt from my parka and used the high prow of the huge cedar canoe to vault to the nearest tree. While I sat down and mused, she frolicked in the branches above.

The canoe lay in an ocean of moss-covered stumps cushioned by undergrowth. In my imagination I saw how the Haida translated the might of a cedar into the might of a war canoe, how they tested its soundness by cutting holes in the trunk, felled it with jade chisels and mallets, stripped the bark, hewed out the hull with adzes and chisels, and finally cut it free from its top. My musings were sharply interrupted by a thump on my shoulders as Gypsy jumped from the tree to the sun-stippled canoe.

That night in the community hall of the Skidegate Indian band, Gypsy didn't tear as madly through the enthralled audience as she had in Haida. While the film was running, she contented herself by sitting on the laps of those in the back row and being petted. But she made herself memorable again when a Haida tot, hardly four years old, passed by her on the way to the washroom. She must have come in late and not been aware of a gibbon in the hall. Suddenly she saw the little black ape and let out a blood-curdling scream. It was sheer horror at the unexpected. It startled everybody in the audience. They jumped up in their seats and turned around to see an innocent and sleepy little gibbon cuddled on one of the mothers' laps. Despite herself, Gypsy had done it again!

But the excitement didn't end. After the show in Skidegate we were kidnapped by a couple of dozen people from Sandspit. They collected our projector, film and suitcase and whisked us away in a yellow crummy owned by the local logging company. Out of the crummy, and into a fast launch, out of the launch and into a car, out of one car and into another. Our friendly kidnappers dumped us into the Sandspit teacherage where we spent the night with the school principal and his wife–in their bed!

Next morning Gypsy and I found that the price of our ransom was to show the film to a quickly organized assembly of school children, housewives and other villagers. Being kidnapped was not so tiring for Gypsy as hiking in the forest, so she was once again an armed bombshell. At first I held her tightly, determined not to let her be a disturbance. But two hours is a long time for a volatile firecracker like Gypsy, so I let her brachiate from the coat racks. But she had learned a new game and intended to play it. While I grasped air she nimbly evaded my far-flung arms, ran down the centre aisle to the table in front of the screen and made her shadow pictures against the film. She chose the most tender moment in the film for her comic relief. I crawled down the aisle below screen level, grabbed her flying feet and dumped her in a small suitcase donated by the wife of the principal and then sat on it.

The hatbox that had brought her to the Charlottes suffered a slow demise. In the stimulating atmosphere of the Misty Islands she had burst from its roof, poked through its doors and pried her way out of its windows too many times for the health of the hatbox. She flew back to Island View in a borrowed suitcase with holes punched in the bottom.

15
And Sweetly She Will Come to Mind

A gibbon's home in the wild is probably a 45-acre treetop lot. Gypsy's home was wherever she went: in the city among teachers and children; at Island View on the seashore among cougars, bears, coatimundis and a macaw; on the sea in a dinghy; on the road in a truck; in newspaper offices and television studios. It seemed a mark of her intelligence that she adapted so well to environments foreign to her heredity.

Yet the few researchers who have tried to test primate intelligence agree that there is a considerable gulf between the intelligence of gibbons and the rest of the great apes. They say that the gibbon, though anatomically more similar to the chimp, orang and gorilla, is more like the monkey in its response to manipulating objects or getting through mazes to reach a tempting food supply.

Most observers say that gibbons are curious and adaptable, as all apes are, but that each species is talented in particular directions. The extroverted chimp excels in the use of implements. The little slower but more persistent orang shows a particular flair for mechanical contraptions. The introverted gorilla seems to be less implement-minded but shows greater persistence that sometimes lets it succeed where the others have failed. The most they'll say for the gibbon is that it is more persistent and persevering than many of the monkeys.

Gibbons do comparatively poorly in intelligence tests. This may be as much the fault of the testing apparatus as the intellectual deficiency of the species. Intelligence tests devised for humans and applied to apes have little meaning. There are different types of intelligence: enemy-prey intelligence, tool intelligence, social intelligence. A gibbon is often distracted by little things close at hand–objects on the outside of the testing apparatus which may be irrelevant to the test itself. Yet it is a mark of intelligence for a gibbon as an arboreal animal to pay close attention to small objects at close range. Then again, many intelligence tests require complicated manipulations of instruments that are difficult for a gibbon's hand to carry out. In addition, gibbons are unsuited psychologically for some tests. They are not highly motivated to work for food rewards, they are relatively unresponsive to shocks and they are easily distracted.

Individual gibbons do show considerable intelligence in using or making tools. Gabrielle, a three-year-old captive at a primate research centre, placed a blanket under a tap so it trapped water and she threaded a rope over a bar to provide a swinging apparatus.

I never gave Gypsy any formal intelligence tests that required her to earn food by solving problems with objects and mazes or manipulating tools. Yet I felt she reached a higher level of intelligence while living with us in her wild-free environment than apes living in the restricted environment of a laboratory cage or even apes in the wild. Gypsy's basic needs were taken care of by humans. She had the time and the opportunity to let her brain develop on higher levels than that of simple subsistence. She had many more social contacts and varied opportunities for play with both people and objects.

Some of her skills were innate. She brachiated, vocalized,

groomed, fed and drank like her wild relatives. Yet by imitation she learned to walk upright with her hands held to her sides, to copy the cries of her playmates, to eat a variety of civilized foods, to drink out of a cup occasionally, to pull down the sheets before settling into a bed, to use sweaters as weapons. She showed insight in the way she learned to get out of enclosures, to open cupboards, to sense she was about to lose her mother when there was no obvious evidence. She showed flexibility in coping with change and challenge, as when she repressed her fear of the unknown by running into strange places on Fire Drill Day, and when she chose to get drenched in Hilary's shower rather than staying in an adjacent room with someone she disliked. She remembered where particular foods were kept, she stole them when she thought she was not being observed, she knew how to manipulate taps when she was thirsty. She adapted to houses, schools and theatres, to cars, buses, planes and ferries. She learned to live successfully with people and it was perhaps people who eventually let her down.

In the third winter of Gypsy's life I settled down at the typewriter in the living room, ready at last to finish the book I had tried to begin over a year before. Susan had left school and was working permanently with David on the film tours. Robin Bennett was helping with a fledgling book-publishing venture. Anne Sinclair, a dynamic, good-humoured lady whose sprightly manner belied her seventy years, was looking after the house and the animals more capably than any assistant before. Scarlett the macaw was getting all the attention she craved. Bubu, who had grown into a strong, healthy bear with an unusually sweet temperament, was given a large enclosure next to Windi and Pixie and Pete. Sam the seal and the other animals in the compound were well and content. Two raccoons, Rocky and Little Rock, lived free outside till they tried to compete with the coatimundis by adding falcon eggs to their diet. Then their life style became somewhat restricted except under supervision while walking with Anne or Susan along the beach.

My only interruption was Chimo, a young cougar newly arrived from the Portland Zoo, who lived in the house while recovering from a succession of sprained and broken legs. Chimo had a history of rickets and a congenital condition of brittle bones. She wore a snowshoe type of splint on her back leg, but none of these factors seemed to hamper her manoeuvrability. She couldn't climb very well so she was able to spend most of her convalescent period in the library and den, which hastened her recovery. When she did come into the living room and in contact with Gypsy, she was able to chase the gibbon, even catching her once by the seat of her pants, but Gypsy always managed to spring above her head and out of the way. To avoid any confrontation with Scarlett, Anne carried the macaw on her shoulder while she did the housework or hung her on the clothesline in the porch.

At night Chimo and Gypsy shared the pen in the living room. Gypsy stayed on the top shelf and swung from the ropes near the ceiling. Chimo, hampered by her snowshoe splint, remained at floor level. Gypsy, who could seldom resist an opportunity to tease, would sometimes dive down to pull the cougar's tail. A large enclosure with a tree trunk and a log cabin was being prepared for Chimo in the compound, and Gypsy would soon be living alone with me in the living room. That was what she seemed to prefer. As the gibbon grew older she increased her resentment of people who touched me by shaking her head and arms at them or grabbing their glasses. When Chimo moved outside, Gypsy would be able to spend more time in sitting on my lap at the typewriter as she loved to do.

Christmas that year was a merry one. There were the usual problems of too much snow and too little water, but with more people to share in the work and an international publisher accepting my first chapters and outline, it seemed I would soon be able to keep my New Year's Resolution and Finish That Book.

But it was to be no Happy New Year. My diary tells the story.

The first week:
I notice that Gypsy's left leg is twitching. She shrieks when I touch it. She tries to hang from her arms but she falls. Crawling on two arms and one leg with her other leg dangling helplessly, she makes pathetic attempts to reach me. She depends on me for her security but I do not know what's wrong.

Has Chimo hurt her? I find no scratches. Has she got her foot entangled in the wire and wrenched a muscle or broken a bone? I check carefully but can find nothing wrong. Has she developed something as a result of competing with too many visitors over the holiday period and then being left alone for a day while we ourselves went out to visit? There are many cases of apes and monkeys dying of psychosomatic diseases.

I phone the doctor's office but to my astonishment his nurse tells me to take Gypsy to the vet. There is no time to argue. I choose the best-known one in town. He x-rays the bones and confirms that none is sprained or broken. He says there is nothing wrong and suggests merely warmth, rest and exercise of the leg muscles. I stare at him in disbelief.

Suddenly Gypsy bursts out shaking as if in an epileptic fit and makes pitiful attempts to reach me across the examining table. David contemplates bruised nerves, trouble in the spinal cord or the brain. The vet laughs at our concern and says such a diagnosis is exaggerated, that Gypsy is just frightened by being in a strange place full of noisy cats and dogs. Gypsy?

We ask the vet to take a blood sample or x-ray the spine. He replies that this is not necessary. We leave.

Pathos is a gibbon unable to move. But to whom do you turn? The vet doesn't know and the medical doctor doesn't want to know. I check out the possibilities myself.

Does she have polio? There hasn't been a case of polio in

Victoria for the past ten years. The clinic is closed and nobody seems to know much about such a disease of the past. To help a gibbon? Is that some sort of monkey?

Gypsy recently raided the medicine chest. Has she taken pills? The druggist tells me she would have reacted more quickly and already have expelled the pills from her system.

Has she licked rat poison? I phone the Poison Control Centre. "How old is your youngster?"

"She's really a two-and-a-half-year-old ape."

"I'm sorry, madam, but we're very busy tonight and we don't give advice for animals. Goodnight." The click is decisive.

Meanwhile muscular tremors penetrate every part of Gypsy's body. She becomes a pulsating mass of vibrations. She shakes her limbs like a zombie. Her head stretches backwards and her eyes plead for the help I am unable to give.

I memorize a chapter in Merck's Veterinary Manual on diseases of the central nervous system. Gypsy's symptoms suggest an inflammation of the nervous tissue of the brain, perhaps meningitis or encephalitis. The viruses that cause these diseases can be picked up by contact with an infected human, may be incubated for weeks, even months, and then take root when the patient's resistance is low or he is subject to a period of socio-environmental upheaval. In man it is usually fatal. Then I read: "Treatment is usually of little value."

I tear another book from the library shelf. "The anatomy of the great apes is so close to that of humans that a medical doctor is much better equipped to treat them than veterinarians." In desperation I phone the doctor again. For over two years he has enjoyed the good times. Why doesn't he help in the bad times? He pays no attention to my pleas. His nurse suggests I take Gypsy back to the Tacoma Zoo. Animals must have health permits to cross borders. Shall I try to smuggle a sick ape into a foreign country?

David goes away again, this time to hospital for an operation on his elbow. His father follows soon afterwards for treatment of his asthma. Both are given continuous care and the correct procedures are set in motion for their recovery.

But no one will help Gypsy. Her time is running out and I don't know what to do. Please God, will You make her well? Or have I no right to ask? I throw myself down beside her and try to will her to live. "Crying is a sign of immaturity," says Anne primly. "It's life." Through my tears I reply curtly, "It's death."

Hilary sends a card specially designed for Gypsy. It reads: "So sorry that you are feeling low but hope you will soon be back in the swing."

I sit at the typewriter and try to write a comedy but it is difficult when I live a tragedy.

The second week:
Gypsy won't eat. To keep up her strength to combat the yet-unknown disease, I feed her strained baby foods. To prevent dehydration, I inject sugar and water under her skin. I give her penicillin antibiotics and increase the heat. Her body temperature is 103°F, four degrees above normal and rising. She shivers.

I phone the pathologist. He is sympathetic and responds to the challenge. Gypsy couldn't have swallowed rat poison because she's not bleeding. She can't have tetanus because her spasms are not yet violent enough. She can't have polio, as with recent preventive vaccine it is almost an unknown disease. He asks me to bring in a sample of her blood and urine.

I take Gypsy to another vet, one fresh out of veterinary school with experience in cats and dogs and horses. "Yes, something appears definitely wrong. I dislike monkeys but I'll give it a try."

He agrees with my diagnosis and treatment. He agrees to another x-ray but finds only congestion in the lungs. I note her harsh breathing, the emission of mucus from her mouth and read that pneumonia and tuberculosis are the most common gibbon diseases.

At last the samples are analyzed and disease is officially confirmed. Infection is rampant. Spreading white blood cells are fighting a losing battle.

I rush to the drugstore to get ampicillin and sulfa drugs. The druggist near Monterey School smiles as he hands me the package. "My two young boys will certainly be impressed that their dad has filled a prescription for their school gibbon." I turn away quickly before he sees my eyes.

Back home Gypsy's muscle spasms continue. As I watch her overworked heart and lungs pumping like a dynamo, I wonder how her frail body can tolerate such punishment. I remember Norm Winnick's words, "Love and security are essential to keep a primate alive." For several long days and even longer nights I hold her in my arms. I keep her warm, alter her position regularly to move around the fluids in her chest, maintain the liquid in the vaporizer and try to ease her breathing. Her temperature is 104°F.

At midnight she utters a little cry, her spinal cord bends backwards into a tight circle, her eyes glazed. Frantic, I call the vet and to my surprise he responds to the panic in my voice and within the hour he is at my door. He arrives with an armload of syringes and bottles, prepared to relieve her distress with death.

But she recovers from the seizure, her breathing becomes more normal, her eyes brighter; she even uses one hand to propel a banana into her mouth. The vet marvels at the change and decides not to put her down. I thank him for coming and walk him to the door.

The pathologist comes for his specimen and leaves with his congratulations. My deep conviction that she can survive if only I have sufficient faith has wrought a miracle.

The third week:

The spasms have stopped. The pneumonia is controlled. Her stools contain little red-and-black sand grains which I am told is blood in the urine. They are analyzed but reveal nothing pathogenic. The rectum swells to the size of a donut and is extruded like a sea urchin. She regurgitates her food. The antibiotics are eating the gut flora. The drugs are stopped. An ointment to relieve the rash is applied. Her temperature drops to a little below normal. She stops trying to bite David when he comes close to me. But she bites the spoon. She bites me.

The manual says, "After effects of meningitis . . . menges destroyed so neural control is lost in some parts of the body. Infection kills some tissue, gradually destroys the brain cells that control the vital functions. Even if the body lives, there may be permanent brain damage."

The long fingers of her glove-like hands reach out to me. But I can do nothing more.

It was a child who helped, a child who knew Gypsy well.

Marilyn, once shy and insecure in my grade six classroom, at ease with Gypsy but not with others, now bubbled with confidence and enthusiasm. She was outgoing, energetic and alive. She told me about the doctor who had wrought such changes.

"Let me take Gypsy to her, Mrs. Hancock. She helped me. She can help you."

"Doctors don't help animals, Marilyn." She sensed my frustration.

"Dr. Marsh will."

It was well after midnight when she led me to Margaret Marsh's office in Oak Bay. Despite the lateness of the hour, the doctor was warm and welcoming. I surrendered Gypsy to her and she motioned me to a seat. There was no time for her to explain her methods. They were unconventional but then all others had failed. This doctor cared. And in her gentle, sympathetic hands Gypsy lay calm and filled with a strength I had not seen in weeks.

As I watched I renewed my faith in people and in life. But for Gypsy help came too late.

At noon the following day, after a single contortion and a shrill cry, Gypsy's heart stopped beating.

No longer does she look like a tired old woman. The fight is over. She is at peace.

I feel empty. One moment she is there. The next she is gone. One lives, one dies. One loves, one loses. The cycle is a simple one.

We're all in love with Gypsy.

... Patience! Time will blur the aching sorrow
And one day sweetly she will come to mind
In gently playful guise; and having smiled–
Aware and not denying, you will borrow
From yourself at each new need to bind
With love a helpless creature.
Give death his due
Or else to emptiness be reconciled.

(Joan Ward-Harris)

Epilogue

This book is not for children but for the children in us all. It is as much written by the children who are Gypsy's friends as it is by me.

We may wish that it had started in the treetops of a far-off country; we may dream that it had ended there. But Gypsy lived amongst us and we must tell her story as it happened, for with children, honesty is the only way. It was we who gave her life and with our words she comes to life again to touch our hearts with joy.

For more than two thousand years Chinese artists and philosophers have celebrated their delight in the music and movement of gibbons. In Chinese culture the gibbon is the aristocrat among apes and monkeys, a magical and mysterious link between man and nature. But as the Chinese forests disappeared, so did the gibbon.

For us, Gypsy wrought magic too. Because of our delight in her, we delighted in all gibbons. And as the forests in Southeast Asia disappear, we hope that the magic Gypsy engenders and the permanent home she has now in the hearts of a much wider audience will help to save the homes of gibbons in the wild.

Acknowledgements

I am indebted to:
Norm Winnick, director of Point Defiance Park Zoo in Tacoma, for starting it all; John Isberg (née Hollebone) for her novel *Guess Who Comes to Our School?*; Joan Ward-Harris for her poem and its perspective; Marian Brown, Marion Condrashoff, Barbara Dickson, Wendy Fischer, Jane Gatt, Leila Harris, Stephen Shortt, Sandra Slaght, Kathy Warren, Sandy Wilson, and other children in my classes for their images of Gypsy in verse; Hilary Stewart, Dr. Graeme Conway and his wife, Andrea, for their continual support and their images of Gypsy in film; Dr. Fulton Fisher and Dr. Paul Heyer of Simon Fraser University in Burnaby for their fresh insights, enabling me to see and also understand; Dr. Margaret Marsh, Marion and Marilyn Dowler for their research in radiesthetics and radionics and their patience in helping me understand.

And although unnamed here, I deeply appreciate the many more who helped in other ways: librarians in Victoria, Prince George, Whitehorse, Fort Nelson, Burnaby and Summerland; directors and technicians at zoos in Calgary, Toronto, Vancouver and Winnipeg; fellow travellers along the road, fellow teachers and students in the classroom–and fellow janitors everywhere.

Lyn Hancock
Mill Bay, Vancouver Island, B.C.

IF YOU WANT TO READ SOME OF THE SCIENTIFIC LITERATURE ON GIBBONS, MAY I SUGGEST:

Gibbon and Siamang vols. 1 (1972), 2 (1973), 3 (1974), 4 (1976), edited by Duane Rumbaugh and published by S. Karger, Basel, Switzerland.

and

Berkson, G. and B. A. Ross. 1968. *Factors to Consider in a Conservation Program for Gibbons* in Behaviour. Proc. 2nd. Int. Congr. Primat. Atlanta, Ga., 1968. vol. 1: 43-47. S. Karger, Basel. 1969.

Bourne, Geoffrey, H. 1977. *Progress in Ape Research*. Academic Press, Inc. New York. 300pp.

Carpenter, C. R. 1940. *A Field Study in Siam of the Behaviour and Social Relations of the Gibbon (Hylobates lar)*. Comp. Psychol. Monogr. 16: 1-212.

Chivers, David John. 1974. *The Siamang in Malaya. A Field Study of a Primate in a Tropical Rain Forest*. S. Karger, Basel/New York. 335pp.

————————— 1977. The Lesser Apes in *Primate Conservation*. ed. His Serene Highness Prince Rainier III and Geoffrey H. Bourne. Academic Press, New York/San Francisco. pp. 539-598.

Gulik, R. H. Van. 1967. *The Gibbon in China*. An essay in Chinese Animal Lore. Brill, Leiden.

Reynolds, Vernon. 1967. *The Apes*. E. P. Dutton and Co., New York.

Rumbaugh, D. M. and Carol McCormack. 1968. *Attentional Skills of Great Apes compared with those of Gibbons and Squirrel Monkeys* in Behaviour: Proc. 2nd. Int. Congr. Primat. Atlanta, Ga., 1968. vol. 1: 167-172. S. Karger, Basel. 1969.

Yerkes, Robert M. and Ada W. Yerkes. 1929. *The Great Apes*. Yale University Press. 591pp.